Leigh S. Cauman
First-order Logic

For John
with my love,

Leigh

Leigh S. Cauman

First-order Logic

An Introduction

Walter de Gruyter · Berlin · New York
1998

∞ Printed on acid-free paper which falls within the guidelines of the ANSI to ensure permanence and durability.

Library of Congress Cataloging-in-Publication Data

Cauman, Leigh S.
First-order logic: an introduction / Leigh S. Cauman.
p.   cm.
Includes bibliographical references and index.
ISBN 3-11-015766-7 (pbk.: alk. paper)
1. First-order logic. I. Title.
BC 128.C38   1998
160--dc21

Die Deutsche Bibliothek – CIP-Einheitsaufnahme

Cauman, Leigh S.:
First-order logic: an introduction / Leigh S. Cauman. – Berlin;
New York: de Gruyter, 1998
ISBN 3-11-015766-7

© Copyright 1998 by Walter de Gruyter GmbH & Co., D-10785 Berlin. –
All rights reserved, including those of translation into foreign languages. No part of this book may be reproduced or transmitted in any form or by any means, electronic or mechanical, including photocopy, recording, or any information storage or retrieval system, without permission in writing from the publisher.
Printed in Germany
Typesetting: Dörlemann Satz, Lemförde
Printing and binding: Werner Hildebrand, Berlin

Preface

Everything in this book I learned either from my teachers, mainly Vevia Blair, who taught me mathematics at the Horace Mann School for Girls, Paul Weiss, who introduced me to logic and philosophy at Bryn Mawr, W.V. Quine, under whom I did my graduate work at Radcliffe, and James Thomson, who used to talk with me about logic and logic teaching when he was at Columbia; or from my books, a world opened up to me by my father; or from my students, who helped me to enjoy profoundly my twenty-five years of teaching and who encouraged me to try to summarize what I have learned. A number of graduate students who taught with me and undergraduates who took our courses used the book while it was being revised for publication; particularly helpful in the work of revision were Jeffrey Barrett, John Bolender, James Murray, and Floyde Bodden. Rivka Kfia has helped me both in proof-reading and in checking deductions and diagrams. Ti-Grace Atkinson has used the book (in typescript) in her teaching and has given me invaluable help and advice in the final work of revision for publication.

The structure of the book follows that of Quine's *Methods of Logic*. I designed it that way not only because that structure seemed right for my purposes, but also so that the book might be usable as an introduction to *Methods*. The exercises are based upon problems designed by many authors, among them Quine, John Cooley, Richard Jeffrey, and Lewis Carroll. The book is intended to introduce intelligent men and women to the principles and notation of modern symbolic logic and to enable them to use those principles and to pursue that discipline elsewhere.

Leigh S. Cauman

Contents

Introduction 1

Part I: Truth-functional Logic
Chapter 1: Principles of Inference 11
Chapter 2: Truth Tables and Truth Trees 73
Chapter 3: Evaluating Arguments 110

Part II: Predicate Logic
Chapter 4: Principles of Inference 145
Chapter 5: Truth Trees for Predicate Logic 187

Part III: Relational Logic
Chapter 6: New Restrictions on the Rules 229
Chapter 7: Truth Trees for Relational Logic 257

Part IV: Identity and Description
Chapter 8: The Logic of Identity 277
Chapter 9: Definite Descriptions 302

Afterword: On Names and Variables 331

Index 337

Introduction

This is an elementary logic book designed for people who have no technical familiarity with modern logic but who have been reasoning, thoughtfully, for years. Such people have learned from experience that even the best of us make mistakes. We make mistakes in reasoning even if we are naturally intelligent and strongly motivated to think straight. So we need systematic procedures that will help us to reason reliably, and we need practice in following those procedures. In other words, there is practical as well as theoretical reason for the study of logic: to train the mind.

Thoughtful students have also learned that there is a significant difference between collecting facts and using them, between absorbing data and, on the basis of these data, building a coherent view of the world (or of some small segment of it). This difference is the ground for the distinction that philosophers draw between truth and validity: truth underlies the reliability of data; validity underlies the reliability of reasoning. It is our intention to reason reliably, in such a way as to preserve truth, that is, in such a way as to be certain that we never derive from premises that are true any conclusion that is not true. So first-order logic, dealing as it does directly with logical practice, is concerned primarily with validity. It sets aside questions about the acceptability of data.

It also sets aside such important issues as how much can be accomplished by the methods of first-order logic, what cannot be so accomplished, and why. These issues are the province of *metalogic*. It is my view that metalogic can be better understood *after* its subject matter (i.e., first-order logic) has been mastered in a systematic, self-conscious way.

First-order logic is the logic of everyday reasoning – formalized, criticized, systematized, and made self-conscious by logicians over the centuries, but nonetheless a tool that we all use in our daily lives. Its basic notions are familiar even to the most unsophisticated readers, precisely because they use them every day. What will be unfamiliar is the focus on reasoning, rather than on what we are reasoning about, and on the structure of that reasoning. Focusing on structure has led logicians to use symbols and diagrams, which some students find daunting, at first. It also tends to make a beginning student uncomfortably self-conscious about his ordinary thinking. Similarly, a student of anatomy, learning about the musculature of his legs and the workings of his nervous system, may become self-conscious about the commonplace procedure of walking.

To the surprise of many beginners, studying first-order logic is an exercise in imagination, disciplined imagination within which it is possible to discover how to reason in a systematic and reliable way. We start from suppositions – not beliefs, not truths, but a species of make-believe – *premises* taken for the sake of argument, in order to see where they lead. From these premises we build arguments, bits of reasoning to support conclusions in which we are interested or which we think are somehow connected to our starting points; and we examine arguments that others have invented.

We *decide* which suppositions we wish to investigate. We *choose* these suppositions. Occasionally we do so randomly, just for fun, to explore the paths they open to us. Many children love to do this sort of thing, being talented and prolific in make-believe. Some mathematicians are geniuses at it, spinning elaborate systems from quite innocent-looking beginnings. But ordinarily we choose our premises with limited objectives in mind: in science, to find the consequences of a hypothesis, which can then be tested against observation or experiment; in decision making, to determine the results of various courses of action, which can then be considered in deciding what course of action to take; in debate, to rebut the views of someone with whom we disagree. In first-order logic we are practicing, in preparation for such genuine reasoning, and we choose our premises in order to get started.

We are free to take any premises we please; for a supposition, unlike a claim or an assertion, cannot be *mistaken* – it is only a starting point for investigation. Nor can it be disallowed on logical grounds, even if it is self-contradictory or frivolous or false in some very obvious way – to disallow it would amount to a refusal to investigate, a closing of the mind.

A supposition can, however, be *pointless*: irrelevant or inadequate to the task at hand. Our work may show us that it leads nowhere, or at least that it cannot be used to demonstrate what we intended to show.

An *argument* consists of a premise (or a number of premises taken together) and a conclusion, which is said to follow. We shall be criticizing (and constructing) arguments of various levels of complexity, and, in so doing, we shall be interested in discovering whether the conclusions do indeed follow from the premises, that is, in discovering whether the arguments are *valid*. The word 'argument' as we shall use it – and as it is ordinarily used in logic and philosophy – denotes not an altercation between conflicting parties, but a bit of reasoning. The term 'valid argument' as we shall use it is a term of approval; it denotes an argument that is "correct" or "logically impeccable," "reliable" in that it cannot be used to lead us from truth to falsity.

In criticizing an argument we shall be questioning the *logical* claim that it makes, not any factual claims made by its component statements (i.e., its premises and conclusion). We shall be questioning whether, on the supposition that the premises are true, the truth of the conclusion is guaranteed. We shall, in other words, be serious about the content of the premises (though without commitment to them); we shall be trying to see what the premises amount to, either in isolation or in the explicit context in which they are proposed. In so doing we will need to make a leap of imagination, such as we must always make in dealing with any serious question.

We shall also be criticizing statements: discovering which among them can properly be classified as *theorems*. A *theorem* is a statement that has been shown to be true, regardless of the facts of the case. There is no set of circumstances under which it could turn out to be false. (If it rains, it rains. If you sing and I dance, I dance. Either the sun is shining or the sun is not shining.) A theorem is *valid* in that it can be counted on to be true; there is no way in which it could *not* be true. It has been shown to be true without reliance upon factual information, on the basis of a valid argument. A theorem summarizes a bit of logical "information." We shall be dealing with arguments (and theorems) in two different but coordinate ways: by testing them to find out whether or not they are valid, and, if we think they are valid, by demonstrating that their conclusions do indeed follow from their premises. We shall use truth trees (see especially chapters 2 and 5) to test both arguments and proposed theorems for validity; and we shall use a system of natural deduction (see especially chapters 1, 4, and 6) to provide a step-by-step demonstration of validity for any argument or theorem we have reason to believe is correct. We shall use these two methods in tandem. (That the two methods agree I take for granted, but to argue the question would be beyond the scope of this work.)

Truth trees provide an indirect, diagrammatic method of testing either arguments or theorems. An argument is valid if there is no set of circumstances in which its premises could be true and its conclusion false. Its truth tree diagrams the various ways in which the premises, taken together with the negation of the conclusion, could turn out to be true. If all the various ways – or *branches* of the tree – are blocked, i.e., shown to be unacceptable, because each of them contains one contradiction or another, then there is *no* way in which the premises could be true but the conclusion false; so the argument is valid. But if there is even one branch unblocked – one loophole – the argument is invalid. The ways in which the premises could be true but the conclusion false can be read off from the truth-tree diagram, and we can see why the argument does not work. Similarly, the truth tree for a purported theorem diagrams the various ways in which that "theorem" could turn out to be false. If even one branch of the truth-tree diagram represents a genuine logical possibility – if it contains no contradiction – the "theorem" fails; it is invalid. Again, the reason for its failure can be read off from the truth-tree diagram.

Truth trees, both with respect to arguments and with respect to purported theorems, help us to understand cases of *invalidity*.

Deductions, on the other hand, clarify cases of *validity*. A natural deduction shows, step by step, why a valid argument works. In detail, using a minimum of basic rules of inference, each rule intuitively acceptable and agreed upon as a basis for work, a deduction shows how one can start from a premise or set of premises and derive a conclusion in which one is interested. It makes explicit the reasoning process of an intelligent human mind.

Because constructing a deduction is a goal-directed activity, dependent upon human ingenuity for its success, this method is more difficult to master than any routine testing procedure. It is also more important. Therefore I have put it first. After having studied the chapters on testing procedures, the reader may wish to go back to chapter 1, in order to understand both methods more clearly.

One concept of particular importance with respect to deduction is that of *discharging* a supposition. We use this first in connection with the principle of *conditional proof.* When some logical work has been done and, on the basis of a premise (call it "such and such"), either alone or in context, a given conclusion (call it "so and so") has been reached, we have *not* established that conclusion. We have instead established the "corresponding conditional": *if* the premises then the conclusion; *if* such and such then so and so. We look back on what we have accomplished and summarize it, drawing the *summary* conclusion: if such and such then so and so. Now we no longer need the supposition from which we started. The summary conclusion holds without the supposition, regardless of whether that supposition was true or not. So we *discharge* the premise. We set it aside, along with the steps of reasoning which followed from it and which we used in reaching our original conclusion -- like scaffolding discarded from a completed building. We do not suppose the premise any longer; we have seen where it leads.

This process should be familiar to the reader from her own reasoning and, in particular, from high school geometry. A related process may also be familiar: what geometers call proof "by cases" and logicians call "dilemma." We start from a number of alternatives (or cases), set forth in a disjunctive statement of the form "p or q or r or ...," a statement which will *not* be discharged. Then we suppose each disjunct separately, investigate the consequences of each of them separately, and, when each has been shown to yield the same conclusion, we discharge the several suppositions together and conclude, over all, that what has been shown to follow from each disjunct individually follows likewise from the original disjunction. (Imagine arriving at a two-pronged fork in a road. We are told that the left-hand fork, the high road through beautiful mountain scenery, leads to Rome, and we are told that the right-hand road, less picturesque but less difficult, also leads to Rome. We have found out that, whichever fork we take, either way, we can get to Rome.)

Similarly, we discharge the supposition in an indirect (or *reductio ad absurdum*) proof. In reductio we make a supposition for the purpose of showing that it is not true. We assume the negation of what we hope to show and try to deduce a logically unacceptable conclusion: if we succeed, in an orderly and rule-bound way, in deriving a direct contradiction, which cannot possibly be true, we have established that the supposition in question was false. And, in reporting the falsity of that supposition, we discharge it: we have made explicit what it had to teach us.

As will be shown in the chapters on deduction, these three ways of discharging premises correspond to three strategies of proof, which, in turn, serve as reasons for choosing premises. We are, as we have seen, free to take any premises we please; but freedom, in argument as in ethics, implies responsibility. When we take a premise we should have a project in mind: to show something or other, and a plan, at least in outline, as to how that demonstration can be accomplished. Designing an informal argument or a formal proof is a purposeful activity, like building a house, preparing a meal, or repairing a washing machine. And the purpose of setting out one's premises is to launch that activity in a forward-looking and reasonable way. Likewise, the purpose of taking an auxiliary premise in the course of a deduction is to launch an activity, though a more limited one. Accordingly it will always be wise to question the *purpose* of making a specific supposition (though we do not question the legitimacy of so doing).

A valid argument starts from premises that mesh reasonably well together and yields a conclusion that could in no way be false provided that the premises are true. An invalid argument does not meet that standard: its premises leave open the possibility that its conclusion could be false. The premises of an invalid argument do not guarantee that the conclusion is false; the argument cannot be corrected by negating the conclusion. This is perhaps surprising. After all, if a third-grader figures that 7 × 8 = 54, he can figure it out again and correct himself: the right answer is 56. Arguments, however, do not work like that: some premises are simply inadequate to yield yes-or-no answers to some questions. So the discovery of invalidity is more likely to show the inadequacy of the premises, which have turned out to be inconclusive, than to show the incorrectness of the conclusions drawn from them.

One way or another, we should be able to understand why an argument is invalid if it is invalid and why an argument is valid if it is valid. It is my hope that the methods and explanations presented in this book will help the reader to recognize an invalid – or a valid – argument when she sees one and to construct orderly and reliable arguments of her own.

Part I
Truth-functional Logic

Chapter 1
Principles of Inference

First-order logic is concerned with rules of reasoning, with providing and justifying such rules, and with teaching the student to use them skillfully and reliably. I shall assume that it is unreasonable to ask anyone to rely upon a set of rules and to use them in work, without first convincing that person that the rules make sense. So I shall start by trying to convince the reader that the rules we will be working with – rules that logicians, mathematicians, philosophers, and scientists have been using for centuries – do make sense.

But, even before that, it will be necessary to convince the reader that it makes sense to use rules at all, to rely upon *some* set of principles in doing intellectual work, rather than to depend only upon the strong (and usually reliable) intellectual intuitions which we all share and to which I will have ultimately to appeal in order to justify the rules of logic. After all, the intellectual intuitions of even the best of us are occasionally unreliable; so all of us may benefit from orderly and critical systems of work.

Reasoning is something we all do. We have a piece of information, or several, and we hold some view or other, and we move from those (premises) to some other view – usually one more concise than what we started from or more illuminating or just more useful for our purposes – which we think follows from them. We move (in mind) from *premises* to *conclusions*. This procedure, which is the primary way in which first-order logic is used, is valuable to us provided that the conclusions we reach are no less certain than the premises from which we started.

Reasoning also occurs in two other sorts of context. We argue *hypothetically* when we are not at all sure of our premises, but want to see where they lead: to test a hypothesis in science, to investigate the wisdom of a possible course of action, or, sometimes, just to practice or to play. Again, the reasoning will be valuable to us only if the connections it articulates are strong, if the conclusions drawn do follow from the premises.

And we reason *indirectly* when we wish to disprove a view that seems wrong. We consider, or "suppose," a questionable view "for the sake of argument" and try to show that it leads to unacceptable conclusions. These conclusions may be inherently unacceptable because self-contradictory, or they may conflict with views generally accepted or jointly held by us and our interlocutors – those with whom we are arguing – and so lead to contradiction in context. Such reasoning serves to discredit the views from which contradiction has been derived. But again, the argument is valuable to us – it achieves its objective – only if the conclusions reached have been shown to rest reliably upon their premises, if the reasoning used is tight and intellectually conservative, that is, if it is not possible that the premises be true but the conclusions false.

How can this kind of conservation be guaranteed?

We note, platitudinously, that it can*not* be guaranteed; since we are human, we can only do our best. But, more importantly, we note that, very often, such conservation is not even desirable. In science, in engineering, in daily life, we often wish to make educated guesses, to move creatively from reasonably secure premises to conclusions that might well be false even if those premises are true. This "inductive" process – reasoning according to the logic of probability – should, it seems to me, also be rule-bound – but bound by rules quite other than the rules we shall be studying. (We shall be studying *deductive* logic.) The point I am making now, however, is that, when we move nondeductively from premises to conclusions that might perhaps be false even if the premises are true, we want to know what we are doing. We are taking risks. And risks should be undertaken only purposefully and with awareness.

First-order deductive logic is not the logic of risk taking. It is the logic of safety, of conservation. It turns out also to be deeply useful in the logic of induction and probability, in part because it helps the investigator to see what risks he is taking.

Granted, then, that we will be restricting our concern to reasoning that is intended to be safe, secure, conservative – why the focus on *rules*?

An *argument*, the mechanism by which we move from one position to another, is recognized to be reliable by virtue of its *structure*. If an argument is reliable – we shall call it *valid* – and another argument is "built" exactly like it, but deals with different subject matter, the second argument is exactly as reliable as the first. This much is generally accepted. We often reinforce our arguments by pointing to other arguments having the same structure (but dealing perhaps with less controversial subject matter) which our interlocutor accepts. We often try to discredit the arguments of our interlocutors by pointing out their similarity to other arguments that are less tempting. In such shared discussions we join in accepting the view that the structure of the arguments is the key to their reliability.

It may very well be that the logician's reliance upon structure is a matter of faith – faith in reason, in rationality. The history of science and technology bears witness to the fact that this reliance upon reason has borne fruit – but we do not wish to use this fruitfulness as evidence for the value of reasoning. For our purposes, what we need to recognize is that reliance on some sort of structure is what rationality *is*. And we shall be trying to be rational.

Remember that an argument is a bridge on which we move from a premise or a set of premises to a conclusion. When we criticize an argument we are not criticizing its premises – the points from which we started. We are evaluating the bridge that takes us from there to here. Our project is to discern the effective structure of that bridge, to articulate and evaluate it. And this we do by means of *rules of inference*. An argument is acceptable – *valid* – if it operates by means of rules that cannot lead us from truth to falsity. An argument is *invalid* if it is possible that its premises be true but its conclusion false. So our rules of inference must be so designed as to be reliably conservative in this sense: no argument constructed in accordance with any of them can start from true premises and yield a false conclusion.

It is desirable that the system of rules we use be not only dependable, as we have seen, but also sufficient for our purposes, that they be complete in that any new and reasonable rule could be derived from those already in use. In that case no additional *basic* rules are needed. It is desirable also that we be able to prove that this is so. But such questions are best postponed. After the use of the rules themselves has been mastered, it is time to reflect upon their power and sufficiency.

Let us now take a look at the basic rules of inference. I urge the reader to question these rules right away. They are not arbitrary. They have been used for a very long time, because they have made sense to many reasoners. But if you are to use them, they must make sense to you.

Rules for "If"

We start with the rules that govern the use of "if," the main connective in a conditional statement. We use the symbol '→' for that connection.

A *conditional* is a claim that so and so – the *consequent* – is conditional on such and such – the *antecedent* – or, as we are more likely to put it,

so and so, if such and such
if such and such, so and so
if such and such then so and so
so and so, provided that such and such
etc.

For example:

The ground is wet if it is raining.
If it is raining, the ground is wet.
If it is raining then the ground is wet.
The ground is wet, provided that it is raining.

It is important to recognize right away the radical difference between the conditional statement itself and its consequent standing alone, the difference in this case between saying that the ground is wet if it is raining and saying straightforwardly that the ground is wet. These two statements are easy to confuse, but failure to notice the difference between them is a bit of carelessness that a responsible reasoner can ill afford.

The difference between the antecedent of a conditional and its consequent is also important. This is a matter of logical order – not, as I hope the sentences shown above have already made obvious, a matter of linguistic order. The word 'if' (or any of its synonyms) introduces the antecedent, whether at the beginning or at the end of the sentence. The '→' in a symbolic restatement of that sentence is placed between the antecedent at the left and the consequent at the right. (It may be well, at least at the outset, to "read" the arrow as "yields" or "leads to.") Using the abbreviations 'r' for "it is raining" and 'w' for "the ground is wet," all four of the sentences shown above would be restated symbolically:

$$r \to w$$

The point of a conditional claim is the use to which it may be put. If we have reason to believe that such and such (the *antecedent*, here: "it is raining"), we have reason to believe that so and so (the *consequent*, here: "the ground is wet"); if we suppose that such and such, we likewise suppose that so and so. (Notice that *these* claims are conditionals.) We are justified in reasoning from such and such to so and so, from the antecedent to the consequent.

This is a one-way street. Since the conditional we are considering does not say:

such and such, conditional on so and so

or

if so and so then such and such (if the ground is wet it is raining)

it does not justify our reasoning in the opposite direction, from so and so to such and such. Taken together, the pair of claims:

{ If such and such then so and so.	(If it is raining the ground is wet.)
{ So and so.	(The ground is wet.)

warrants no conclusion whatsoever (other than mere repetition of the claims already made); it tells us nothing at all about whether such and such (whether it is raining), one way or the other. This *asymmetry* of the conditional is something we will have to watch out for, throughout our work.

It may be well to mention here another common source of confusion, which will be discussed later on: the difference between 'if' and 'only if'. Because these expressions are quite different in meaning, failure to notice the appearance of the word 'only' can lead to misunderstanding. 'If' introduces the antecedent of a conditional statement; 'only if' does not.

The first rule we shall be using, then, is *modus ponens*[1]:

Modus Ponens: From a conditional statement, taken together with its antecedent, its consequent follows as a conclusion.

Accordingly,

$$\frac{p \rightarrow q}{\therefore q} \quad \text{MP}$$

where p is any statement whatever, true or false, simple or complicated; q, similarly, is any statement whatever, not necessarily distinct from p; $p \rightarrow q$ is the conditional: if p then q; '$\therefore$' means "therefore"; and the bar separates the premises from the conclusion. The principle of Modus Ponens tells us that, given any conditional statement (say, $p \rightarrow q$) as a premise and, as a second

---

[1] '*Modus Ponens*', a Latin phrase which has somehow persisted, is short for "*modus ponendo ponens*," "way of putting by putting", or "way of affirming by affirming"; the consequent of a conditional is affirmed "by" or "as a result of" affirming the antecedent.
Modus ponens is often referred to in modern parlance as the "rule of detachment"; given the antecedent of a conditional, the consequent may be "detached" from that conditional and affirmed on its own. Also, language-oriented or computer-oriented modern systems often call it the rule of "$\rightarrow$ elimination"; the '$\rightarrow$' that appears in the conditional is "eliminated" in the conclusion.

Principles of Inference 17

premise, the antecedent of that conditional (*p*), it is legitimate to draw as a conclusion the consequent of that conditional (i.e., *q*). The rule Modus Ponens has been stated here in English. The symbolic representation, or "restatement" that follows it is *not* the rule or even part of it; it is a *sample*, helpful for remembering the structure of what the rule conveys. The rule itself is given by the English-language formulation that precedes the symbols.

We shall be using symbols of two sorts: logical symbols (sometimes called "logical constants") for connectives of various kinds, needed to dispel ambiguity (the arrow is one such), and "variables," single letters that abbreviate sentences in Part I or noun phrases or predicate phrases later on. When compound statements (conditionals, for example), that is, statements built from other statements by means of the logical connectives, are expressed in symbols, the resulting expressions will be called "statement forms."

Using these symbols is conducive to brevity and clarity of thought, for it helps us to push aside irrelevant detail and allows us to focus on the structure of the reasoning. Throughout our work we shall be using symbols not for their own sake nor for the sake of constructing an independent symbolic language, but primarily as an intellectual tool. They help us to see the structure of our arguments.

The use of the rule Modus Ponens is unrestricted; it is legitimate in every case, that is, whatever statements in whatever context are put in place of '*p*' and '*q*'. Correct use of Modus Ponens depends, of course, on correct reading of the conditional; '$p \to q$' is not to be confused with '$q \to p$'. The second premise in this form of argument must be *precisely* the antecedent of the conditional being used, and the conclusion drawn must be *precisely* the consequent of that conditional.

Let us consider an example. Suppose we are convinced that if George is in Hartford (*h*) George is in Connecticut (*c*). We learn that George is indeed in Hartford, visiting his sister. We conclude that George is in Connecticut. The argument we are using looks like this:

$$h \to c$$
$$\underline{h}$$
$$\therefore c$$

and is valid, according to the principle of modus ponens.

It should be obvious, on the other hand, that we would be foolish to argue in this way:

$$h \to c$$
$$\underline{c}$$
$$\therefore h$$

That is, if we were convinced that if George is in Hartford he is in Connecticut and we then learned that George is in Connecticut, we would not have been given reason to believe that George is in Hartford. The second "argument" is an instance of the *fallacy of affirming the consequent*, which is the "corresponding fallacy" of modus ponens. (In modus ponens we *affirm the antecedent* as our second premise.) There are occasions when one is tempted to mistake the second pattern of argument for the first.

One further point must be made about the meaning of the conditional before we go on to the second rule. As we shall use it, the conditional claim '$p \to q$', or 'q, if p', or 'so and so, if such and such', contains no indication as to *why* it is said to hold; what is being claimed is *that* it holds. In the "real world," of course, when we have reason to believe a conditional, we (ordinarily at least) have reason to believe that there is a connection of some kind between its antecedent and its consequent; examples abound:

If Mary crosses against the light, Mary is in danger. $(c \to d)$
If Henry is in the kitchen, Henry is in the house. $(k \to h)$
Alice will wear her hat if it rains. $(r \to h)$
etc.

Nonetheless the conditionals themselves do *not* report these connections; the conditionals omit – or abstract from – the reasons behind them. Thus we build a safety factor into our reasoning. We set aside various considerations that might strengthen our confidence in the conclusions we reach, but which we do not need.

In using modus ponens we do not depend upon causal or other connections between antecedent and consequent; we depend only on the fact (or supposition) that, somehow or other, the consequent holds provided the antecedent does. All this despite the fact that, both in philosophy and in social life, the reasons that support a claim are often more important or more interesting to us than the claim itself. In orderly deductive reasoning, the conditionals we work with are drained of causal or purposive or other significance and reduced to bare claims.

We turn now to the second rule for the conditional, the rule of *conditional proof.* In using this rule we do not *use* conditionals; we *establish* them. We establish them on the basis of scientific, historical, logical, or other disciplined intellectual work, in context. Many elementary logic books leave the principle of conditional proof until late in their exposition, or even omit it entirely, to be dealt with in the context of more advanced work. I prefer to take it up right away.

Conditional Proof: The process of deriving a conclusion validly, in a rule-bound way, from a premise (or set of premises) justifies the claim made by the *corresponding* conditional: if the premise(s), then the conclusion.

Accordingly,

$$\begin{array}{c} p \\ \cdot \\ \cdot \\ \cdot \\ \underline{q} \\ \therefore \ p \to q \ \ \text{CP} \end{array}$$

where p is a premise here introduced for purposes of argument and where the dots abbreviate a bit of rule-bound reasoning.

This pattern of argument can occur, as we shall see, either on its own, as in this sample, or in the context of a more complicated argument, where other premises are also at work. The pattern itself will be clearer after the "star system" has been introduced (in the next few pages); the star system will make it visually clear that the premise (here, p) is *discharged* (see Introduction, pp. 5, 6) when the rule CP is used. What results from a correct derivation of a conclusion (q) from a premise (p) is the *summary conclusion* ($p \to q$), that is, the conditional statement corresponding to the derivation.

The sample shown above, with its three vertical dots abbreviating an unspecified bit of "rule-bound" reasoning, may look odd to the reader, especially since he will have noticed that the rules by which the reasoning is bound have not yet been expounded. But they will be.

Also, what allows us to introduce the premise (p)?

But then again, what could disallow it? Many systems, textbooks, computer programs, etc. use something called a "rule of premises," which tells us that it is OK to introduce any premise, of any kind, at any point in any argument. This is of course correct, and it needs to be stated explicitly and used with awareness. But it does not seem to me to be a *rule*. *Of course* we may introduce, or suppose, any premises we like; what is important is that we remember that we have introduced them, that we keep track of them and see where they lead.

The orderly reasoner is free to make any supposition, or take any premise, at any point in the course of the work. There is no question about the "right" to take a premise, any premise, for the sake of argument; what must be questioned is the purpose of so doing. And one such purpose is in order to establish a conditional, by means of the principle of conditional proof. (Other such purposes are for indirect proof, or reductio ad absurdum, and for use in a dilemma; see below.) If, on the supposition that such and such, we are able to prove that so and so, we will have established that if such and such then so and so. If, on the basis of a batch of earlier premises, on the further supposition that such and such, we are able to prove that so and so, we will have shown that, on the basis of those earlier premises, which do not ordinarily include such and such, if such and such then so and so.

At any point in a systematic course of reasoning, a certain number of premises can be seen to be *in play*[2]: we have "taken"

---

[2] The phrase 'in play' is due to Nancy Middleton (note the analogy to various ball games, notably football and tennis). The phrase 'in force' also occurs in the literature, with the same meaning. I have used both 'in play' and, where appropriate, 'available'.

those premises; we are working with them, trying to discover where they lead. (Notationally, we shall make clear which premises are in play at each step of our deductions, by citations to the left of the step number, denoting each premise in play either with a star or – in the special case of the rule DILemma – with an appropriate capital letter, which functions like a star.[3]) The rule of conditional proof is a rule for *discharging* premises.[4]

A bit of valid reasoning moves from premise p to conclusion q; CP tells us that this reasoning warrants the conditional conclusion $p \to q$. Accordingly, p is no longer needed (as a premise); we have seen where it leads and have summarized that information in the conditional '$p \to q$'. The premise p is no longer in play; it is *discharged*. The conditional conclusion stands without it.

Recapitulating, and marking with a star each step in the argument at which the premise p remains in play:

```
*  p    PREM
*  .
*  .
*  .
*  q
   ∴ p → q    CP
```

---

[3] Various diagrammatic methods have been used to make clear which premises are in play at any point in a deduction. See, for example, John M. Anderson and Henry W. Johnstone, Jr.: *Natural Deduction* (Belmont, Calif.: Wadsworth Publishing Company, 1962), pp. 9 ff. See also Merrie Bergmann, James Moor, and Jack Nelson, *The Logic Book* (New York: Random House, 1980), pp. 134 ff. Any unambiguous bookkeeping system that is not too cumbersome will do. Albert E. Blumberg, in *Logic: A First Course* (New York: Alfred A. Knopf, 1976), cited the step numbers of the premises in play, instead of stars, according to a routine rather similar to the star system used here. That system, which is illustrated in the next few pages, derives from W.V. Quine, *Methods of Logic* (New York: Henry Holt and Company, 1950), pp. 153/4. A memo on this star system appears among the Reminders at the end of chapter 1 of this book; see pp. 59–60.

[4] We might call it the "rule of discharge" in analogy to the "rule of detachment" for modus ponens. A more usual term is '→ introduction'.

Note that there is no star at the conclusion; the premise p at which the star was introduced has been discharged.

An example, albeit an abstract one, seems to be in order. We will provide a demonstration of a case of reasoning by the *chain rule* (traditionally called "hypothetical syllogism") – a rule that is undoubtedly more familiar to the reader and more immediately and intuitively acceptable than either MP or CP:

> CHAIN: From a pair (or sequence) of conditional statements in which the consequent of one conditional is the antecedent of the next, the conditional statement that has the first antecedent as its antecedent and the last consequent as its consequent, follows as a conclusion.

To Prove:
$$* \begin{cases} p \to q \\ q \to r \end{cases}$$
$$\therefore \ * \ \overline{p \to r}$$

where the appearance of the star attached to the conclusion makes clear that the two premises, which together constitute our starting point, are *not* discharged.

In the demonstration, the introduction of a star will mark the taking of a premise or a pair or triplet of premises; the dropping of that star will mark its discharge. We start with the pair of premises:

$$* \begin{cases} 1 & p \to q \\ 2 & q \to r \end{cases} \quad \text{PREM}$$

To prove: $p \to r$.

In the context of the premises 1 and 2, an *auxiliary premise p*, the antecedent of the desired conclusion, is taken at step 3 and marked with a second star; then this premise is used, in steps 4 and 5, to derive the conclusion r, the consequent of the desired conclusion. Such auxiliary premises are often needed to initiate subproofs, and the second star (or sometimes a third or fourth star) makes visually clear the extent of the subproof. Each star at each step reminds the reader that a specific premise is still in play.

```
**  3  p            PREM
**  4  q            1,3 MP
**  5  r            2,4 MP
```

Then, at step 6, we summarize the work done in steps 3–5 and report the desired conclusion:

```
 *  6  p → r        3–5 CP
                    QED[5]
```

At step 6, the premise 3 (which was p) is discharged; its star is dropped. So step 6 tells us (since it carries one star, not two) that $p \to r$ follows from the original premises $p \to q$ and $q \to r$, without p. Notice that the annotation to the right of step 6 *says* that step 3 yielded step 5, which is *true:* the subproof really did start from 3, which was p, and arrive at 5, which was r.

I like to call this standard for the correct use of the principle of conditional proof, the standard of "truth in packaging", in analogy to standards of truth in packaging enforced in grocery stores and pharmacies. Use of CP that is fraudulent, i.e., not "truthful" in this way, is fallacious. When the principle of conditional proof is used correctly, the summary conclusion drawn is a conditional whose antecedent is (in fact) the last premise taken and whose consequent is (in fact) a conclusion drawn from that premise in the context of the deduction. The justification for the claim that the conditional holds in that context is the subproof that starts from that premise and yields that conclusion (here: from 3 to 5, written '3–5' with a dash; *not* 3 and 5, which would be written '3,5' with a comma); it is reported, truthfully, in the citation to the right of the step in question. The subproof is summarized and reported; its premise is discharged.

If this abstract little proof is hard to read, try putting ordinary-language sentences for 'p' and 'q' and 'r' – any sentences you like – and rereading it. Here is an example.

---

[5] *Quod erat demonstrandum:* what was to be shown; mission accomplished. We shall use this mark of success at the conclusion of each successful deduction; we shall omit it at the conclusion of "deductions" used to illustrate mistakes in work.

Let p be "Joe is in the kitchen." Let q be "Joe is in the house." Let r be "Joe is in Boston." Let us suppose that

* $\begin{cases} 1 & \text{If Joe is in the kitchen, Joe is in the house.} \\ 2 & \text{If Joe is in the house, Joe is in Boston.} \end{cases}$ PREM

The chain rule will tell us that it follows from these premises that "if Joe is in the kitchen, Joe is in Boston." To show this without (premature) reliance on the chain rule, we suppose further that

** 3 Joe is in the kitchen.   PREM

and we see what follows from that:

** 4 Joe is in the house.   1,3 MP
** 5 Joe is in Boston.   2,4 MP

Finally, using CP, we note that we have shown:

* 6 If Joe is in the kitchen, Joe is in Boston.   3–5 CP
                                                    QED

The single star at step 6 registers the fact that that conclusion depends upon premises 1 and 2, but not upon premise 3. The subproof 3–5, summarized at step 6, can now be set aside, like scaffolding that has served its purpose.

Let us pause to consider what we have accomplished. We have provided a justification for using an *auxiliary rule*, the *chain rule*:

CHAIN: From a pair (or sequence) of conditional statements, in which the consequent of one conditional is (precisely) the antecedent of the next, the conditional that has the first antecedent as its antecedent and, as its consequent, the last consequent, follows as a conclusion.

We have accomplished this by *using* our first two basic principles: Modus Ponens and Conditional Proof, which we did not attempt to establish. Instead we asked the reader to accept those principles as part of our starting point in the discipline of logic. We established the acceptability of the (auxiliary) Chain Rule by doing a deduction for a *sample* argument:

$$p \to q$$
$$q \to r$$
$$\therefore p \to r$$

that is, by deducing its conclusion from its premises. (We could instead have used the longer sample argument:

$$p \to q$$
$$q \to r$$
$$r \to s$$
$$\therefore p \to s$$

or even a longer one, but this seemed unnecessary, since the method would have been the same.) Since the method used here (using only MP and CP) will work likewise for any "chain" argument, no matter how complex and no matter how long, there is no need to spell out these steps again; any chain argument can be similarly derived, using MP and CP only. Accordingly, to save time, energy, and paper, wherever convenient, we feel free hereafter to use the newly established chain rule as a principle of inference.

The Principle of Conditional Proof is not as unfamiliar as it may perhaps have seemed at first. We use it every time we work out in our minds the consequences of a plan of action or a possible turn of events. We take the plan or the happening as a premise, or supposition – set in the context of our other relevant beliefs – and see where it leads. We articulate our view of where it leads (our conclusion) and arrive at a *conditional* statement, with the plan or happening as its antecedent and our conclusion as its consequent, a statement which we are then prepared to defend on the basis of our relevant background beliefs. We use the principle of conditional proof to establish such conditional statements, which are then available for use in other ways.

The principle CP provides us with one *strategy of proof,* the first of many that will be pointed out as we go along. If we have reason to believe that a conditional statement (say, If Joe is in the kitchen, Joe is in Boston) holds in a given context and if we want to prove that this is so, we begin by taking as a premise the antecedent of the conditional in question (Joe is in the kitchen) and, using that premise (in this case an *auxiliary* premise, for there are other, prior ones), we try to prove the consequent (Joe is in Boston). If we are successful and then summarize our work, by CP, we have proved what we set out to prove (here, If Joe is in the kitchen, Joe is in Boston).

The Rule Repeat

I introduce next a principle which is so obvious as to seem unnecessary and which is seldom used explicitly except for convenience in exposition: the rule REPEAT (R):

> REPEAT. In the course of a deduction it is always legitimate to REPEAT a premise that is still in play or a step in proof whose premise is still in play.

This use of a step in proof – whether it is a premise or a step derived from a premise – is legitimate if and only if the step in question is indeed in play (it has not been discharged).

Let us use this principle to derive what the Greeks called the "law of identity" – one of their three "laws of thought": identity, noncontradiction, and excluded middle – the theorem '$p \to p$'.

To Prove: $p \to p$

```
 * 1  p            PREM
 * 2  p            1 REPEAT
   3  p → p        1-2  CP
                   QED
```

A *theorem* in logic, as in geometry, is a statement that has been shown to hold, regardless of fact. It is valid in that there is no set of circumstances in which it could turn out to be false. Its proof contains no undischarged premises. (Occasionally, we hear logicians say that a theorem is a statement "provable without premises" – but this is of course elliptical for "provable without premises that have not been discharged." It would be hard to find a proof without any starting point at all.)

The lack of stars at the last line of this derivation marks that line as a *theorem*: true without dependence upon any premise or supposition, established on logical grounds alone. Instances of the *law of identity* are:

If it rains, it rains.

which is true, regardless of the weather.

If Mary likes carrots, Mary likes carrots.

which is also true, in any circumstances. Etc.

A second theorem may be of interest: $p \to (q \to p)$. Instances are: If it is raining then if I like the rain it is raining; If it is raining then if I don't like the rain it is raining; If it is raining then if today is Thursday it is raining; etc. Of course we all know that if it's raining it's raining whether we like it or not; so this theorem seems intuitively reasonable. Here is a deduction:

To prove: $p \to (q \to p)$

```
 *  1  p                PREM
**  2  q                PREM
**  3  p                1 REPEAT
 *  4  q →p             2-3 CP
    5  p →(q →p)        1-4 CP
                        QED
```

It may be well to reread this deduction in English, indicating as we go along how considerations of strategy help us to design the proof.

To Prove: If it is raining, then if today is Thursday it is raining.

Our basic strategy will be Conditional Proof. Our *last* step – we cannot be sure how many steps we will need – will be, say,

8 If it is raining then if today is Thursday it is raining.

Accordingly, the first step will be to suppose as premise the antecedent of the desired conclusion. So we have:

* 1 It is raining.   PREM

And the second-to-last step, like the first, will be dictated by the strategy of Conditional Proof; it will be the consequent of the desired conclusion. So we have:

```
 *  7  If today is Thursday it is raining.
    8  If it is raining, then if today is Thursday
       it is raining.                                 1-7 CP
```

Now we must derive step 7 – again by Conditional Proof; that is, we assume the antecedent:

```
** 2  Today is Thursday.                              PREM
```

and try to prove the consequent:

```
** 6  It is raining.
```

But that is easy: this statement follows, by REPEAT, from step 1. We say so:

```
** 3  It is raining.                                  1 REPEAT
```

and we adjust the numbering:

```
 *  1  It is raining.                                 PREM
 ** 2  Today is Thursday.                             PREM
 ** 3  It is raining.                                 1 REPEAT
 *  4  If today is Thursday, it is raining.           2-3 CP
    5  If it is raining then if today is
       Thursday it is raining.                        1-4 CP
                                                      QED
```

It is worth noticing that the design of this deduction has been accomplished *from the bottom up*: we began by focusing on what we intended to prove. It is also worth noticing that, like the deduction for the CHAIN rule, this deduction required the use of an auxiliary premise (step 2) marked by a second star; the resulting pair of stars makes evident the extent of the subproof.

We have here illustrated the use of the rule REPEAT. The main point of interest with respect to this rule, however, is not its usefulness, but its limitation, the explicit restriction on its use. A step in proof may be repeated if and only if its premise is still in play.

Violation of that restriction (the "fallacy of illicit repetition") can lead to dreadful confusion. Suppose, for example, that we were to add a line (step 7 or step 7′ below) to our chain-rule derivation:

$$
\begin{array}{lll}
* \left\{ \begin{array}{l} 1 \; p \to q \\ 2 \; q \to r \end{array} \right. & \text{PREM} \\
** \; 3 \; p & \text{PREM} \\
** \; 4 \; q & 1,3 \; \text{MP} \\
** \; 5 \; r & 2,4 \; \text{MP} \\
* \; 6 \; p \to r & 3\text{--}5 \; \text{CP} \\
* \; 7 \; r & 5 \; \text{REPEAT} \\
& (\text{Joe is in Boston})
\end{array}
$$

or

$$
\begin{array}{ll}
* \; 7'\; q & 4 \; \text{REPEAT} \\
& (\text{Joe is in the house})
\end{array}
$$

What nonsense! The notion that either of these "conclusions" (7 or 7′) follows from our premises (1 and 2) is tantamount to the view that what does follow from 1, 2, and 3 together follows from 1 and 2 alone. After step 6, where premise 3 is discharged and the star associated with premise 3 is dropped, steps 3, 4, and 5 are no longer available, whether for repetition or for any other use.

Similarly – and equally fallaciously – one might append to the derivation of the law of identity a fourth line:

$$
\begin{array}{lll}
* \; 1 \; p & \text{PREM} \\
* \; 2 \; p & 1 \; \text{REPEAT} \\
 \; 3 \; p \to p & 1\text{--}2 \; \text{CP} \\
 \; 4 \; p & 2 \; \text{REPEAT} \\
& (\text{or } 1 \; \text{REPEAT})
\end{array}
$$

which would suggest that *any* statement – say, "It is raining" or "Alice likes carrots" – could be demonstrated on logical grounds alone and so could be taken to be a theorem of logic.

A sensible arguer would be most unlikely to make either of these mistakes, but other similar mistakes are often made. The point is that it should always be clear, at every point in any careful argument, which premises (and steps derived from them) are in play and which are not. Only those premises and steps which are in play are available for repetition or other use. The star notation provides one bookkeeping system to help us avoid mistakes. Since the star notation registers the taking and discharging of premises and so makes visually clear which premises are in play and which are not, the fallacy of illicit repetition can be avoided by careful attention to the stars. The injunction: set aside, as having done its work, a subproof whose premise has been discharged, can be rephrased: never cite, as justification for a step in proof, an earlier step that carried any stars other than those appropriate to the step that is being justified.

Rules for "And"

We turn now to the rules for "and", which is written '∧' and understood here as connecting statements – not as connecting properties or things. There are just two such rules: SIMPlification (∧ elimination) and ADJunction (∧ introduction). Both of these rules are easy to understand and accept and use, but both are easy to misuse through carelessness.

A *conjunction* is a compound statement built from other statements by means of the connective "and" or any of its synonyms ("as well as", "besides", "but", etc.); a conjunction jointly asserts all its *conjuncts*, or conjunctive components. To *use* a conjunction we *simplify* it: we separate out a conjunct and set it forth as a step in argument, for convenience in further reasoning. To *establish* a conjunction we bring together its conjuncts, which have been separately supposed to hold or separately proved; we *adjoin* them. I begin with SIMPlification:

> SIMPlification: A conjunctive component of a conjunction follows from the conjunction.

If I know that Mary likes carrots and Henry likes peas, I can be sure that Henry likes peas; likewise, that Mary likes carrots.

Accordingly,

$$\frac{p \wedge q}{\therefore p} \quad \text{SIMP} \qquad \frac{p \wedge q}{\therefore q} \quad \text{SIMP}$$

Unfortunately, the limitations of language, whether written or spoken or conveyed in the sign language of the deaf, require that we illustrate the use of this rule at least twice, to make clear that it does not matter which component of the conjunction we choose to derive from it. It would be helpful, perhaps, to continue:

$$\frac{p \wedge q \wedge r}{\therefore q} \quad \text{etc.}$$

For *any* conjunctive component of a conjunction follows from the conjunction itself. This is, I think, intuitively clear. But language occurs sequentially, whether in time or directionally on the written page (left to right or right to left or up to down, etc.) and so fails to convey with immediacy the radical symmetry (or directionlessness) of the logical "and", which is symbolized by '∧'. We know, of course, that if I say any of

Mary likes carrots, and Henry likes peas.
Joe is poor, but Joe is honest.
Snow is white, and grass is green.

it does not matter in the least which one of each pair of sentences I mention first. But I must mention one first; I cannot say them simultaneously or write them on top of each other and still make myself understood.

So we shall take it that the law of simplification allows us to derive from a conjunction any of its conjunctive components regardless of the order in which these are presented, on the ground that the order of presentation does not matter.

But, an alert reader will object, there are many sentences in which the order of presentation does matter. "Mary saw Henry and ran out of the house" seems to say something quite different from "Mary ran out of the house and saw Henry." Both of these sentences can, of course, be interpreted as conveying more than is conveyed by simple conjunction, and they surely would be so interpreted if they appeared in a novel. (Maybe Mary is afraid of

Henry or, on the other hand, is eager to be with him; maybe Henry has been hiding behind the barn; etc.) For our purposes, however, such differences do not matter. The logical 'and', or '∧', like the logical '→', abstracts from connections, causal and otherwise, between the statements that it brings together. From either of our two sentences we can derive "Mary saw Henry" by simplification; likewise, "Mary ran out of the house." And either of the two sentences will count as true if Mary did both of those things, in whichever order, for whatever reasons.

Conversely, the rule of ADJunction tells us that, given any statements taken separately, we may derive their conjunction.

> ADJunction: From any statements in play at any point in a deduction their conjunction follows as a conclusion.

Accordingly,

$$\begin{array}{c} p \\ \underline{q} \\ \therefore\ p \wedge q \end{array} \text{ADJ} \qquad \begin{array}{c} p \\ \underline{q} \\ \therefore\ q \wedge p \end{array} \text{ADJ} \quad \text{etc.}$$

The first caveat to be noted is that the statements conjoined must be "available" at the point in the argument at which they are conjoined; that is, if they are premises they must be in play, and, if they are not premises, their premises must be in play (see above, the restriction on the rule REPEAT).

Secondly, the conjunction must be built properly. "Either Mary is interested in music and so is Henry or else music is a bore" is not the conjunction of "Either Mary is interested in music or else music is a bore" and "Either Henry is interested in music or else music is a bore", nor is it the conjunction of "Mary is interested in music" and "Either Henry is interested in music or music is a bore"; it is not a conjunction at all. Many sentences with "and" in them are not conjunctions.

The restriction on availability is reflected in the star system. A statement is available for repetition or for other use (in this case, adjunction) if the step at which it is to be used carries at least those stars which were carried by the step cited. The following "deduction" is, accordingly, fallacious:

If Henry is accepted in the program, he will be studying economics and hospital management.

Therefore, Henry will be studying hospital management and if he is accepted in the program he will be studying economics.

$a \to (e \wedge h)$ To Prove: $h \wedge (a \to e)$

```
 *  1  a → (e ∧ h)       PREM
**  2  a                 PREM
**  3  e ∧ h             1, 2 MP
**  4  h                 3 SIMP
**  5  e                 3 SIMP
 *  6  a → e             2–5 CP
 *  7  h ∧ (a → e)       4, 6 ADJ
```

The mistake – the *fallacy of illicit repetition* – occurs at step 7: the conjunct h is derived illicitly from step 4, which carries two stars; its premise (a, step 2) has been discharged at step 6, and so neither a nor h is available at step 7.

Contrast with this invalid argument the valid argument that goes in the opposite direction:

Henry will be studying hospital management and, if he is accepted in the program, he will be studying economics.

Therefore, if Henry is accepted in the program he will be studying economics and hospital management.

$h \wedge (a \to e)$ To Prove: $a \to (e \wedge h)$

```
 *  1  h ∧ (a → e)       PREM
**  2  a                 PREM
**  3  h                 1 SIMP
**  4  a → e             1 SIMP
**  5  e                 4, 2 MP
**  6  e ∧ h             5, 3 ADJ
 *  7  a → (e ∧ h)       2–6 CP
                         QED
```

It may be worth while to look for a moment at the logical structure of these two statements. One of them '$h \land (a \to e)$' is a conjunction and can be simplified (see steps 3 and 4 of this deduction); one of its conjuncts is a conditional. That conditional, '$a \to e$' can be used as a premise for modus ponens (see step 5), but the whole statement '$h \land (a \to e)$' cannot be so used. The other statement '$a \to (e \land h)$' is a conditional with conjunctive consequent; since it is a conditional it cannot be simplified.

Note further that the ordinary-language sentence 'If Henry is accepted in the program he will be studying economics and he will be studying hospital management' is ambiguous as between our two statements; for it can be read as either '$a \to (e \land h)$' or '$(a \to e) \land h$' ['$(a \to e) \land h$' and '$h \land (a \to e)$' come to the same thing]. Logical notation, which insists on the use of parentheses for grouping, so as to make clear precisely what the connectives connect, and which disallows as ill formed such "formulas" as '$a \to e \land h$', which omit needed parentheses, can be useful in forcing us to decide what our ordinary-language sentences were initially intended to convey. (See the memo on parentheses, among the Reminders for this chapter, p. 61.)

The rule ADJ allows us to build a conjunction from its component parts; for the conjunction is true provided its components are. The rule SIMP allows us to separate out any of the components of a conjunction; for the components are true if the conjunction is. This is what 'and' means.

The reader will remember that both Modus Ponens and the Chain Rule began with a pair of premises *taken together*, from which a conclusion could be derived. Other rules of inference – modus tollens, elimination, and nonconjunction (see below) – and many of the arguments we shall analyze and evaluate, start in this way. When we "take" two or more premises "together," we are actually making one supposition: we are supposing the conjunction of those premises. So the notation:

$$*\begin{cases} 1 & p \\ 2 & q \end{cases} \qquad \text{PREM}$$

is short for

```
* 0  p ∧ q          PREM
* 1  p              0 SIMP
* 2  q              0 SIMP
```

Since the conjunctive premise, here nicknamed Premise 0 (Premise zero), is often extraordinarily clumsy, we shall continue to use the first form.

At this point, now that the rules for '→' and '∧' have been set out and the star system explained, the reader may wish to begin doing deductions. Problems II.1 to 4 (see page 69) can be done even before the rules for negation and disjunction are introduced.

Rules for Negation

We turn next to the rules for negation, written '~'. Negation is a basic logical concept with which we are all familiar. Small children usually learn to say "no" long before they learn to say "yes"; and they use "no" correctly: they push away food they do not want to eat or refuse to do something they do not want to do, without suggesting alternatives. More sophisticated people often overestimate the information content of a statement of denial. They mistake a reason for the denial for the content of the denial itself. The denial, or negation, of a statement, written '$\sim p$', is the straightforward claim that what the statement p says is not so; "$\sim p$" has no further content. The denial of "It is snowing" is "It is not snowing" – which tells us nothing else about the weather. And the denial of a complicated statement denies that statement taken as a whole; it does not deny (or otherwise modify) the component parts.

The rules for negation rest on the second of the Greek "laws of thought," the principle of noncontradiction: it is impossible that a self-contradictory statement be true. That is, in general, for any p, $\sim(p \wedge \sim p)$.

It does not seem sensible to me to try to argue in support of this claim; it is correct on the face of it, and any argument in its support will depend upon it. But it may very well be a metaphysical assumption – an assumption integral to the enterprise of coherent thinking, which I shall not attempt to defend.

The first rule of inference for negation is *Reductio ad absurdum* (reduction to absurdity, RED for short, or ~ introduction). The pattern of its use is often called "indirect" proof. It has much the same structure as conditional proof: its starting point is a premise, taken purposefully for the sake of argument; there follows some rule-bound logical work; and the premise is discharged when the rule is invoked.

REDuctio ad absurdum: The process of deriving a direct contradiction, in a rule-bound way, from a premise, justifies the denial of that premise.

Accordingly,

$$\begin{array}{ll} * & p \\ * & . \\ * & . \\ * & . \\ * & \underline{q \wedge \sim q} \\ \therefore & \sim p \qquad \text{RED} \end{array}$$

where q is not necessarily distinct from p.

The process of deriving a direct contradiction, however complex (It is raining and it is not raining; the numbers n and m are both even and it is not the case that the numbers n and m are both even, etc.), legitimately, in a rule-bound way, from a premise, justifies the denial of that premise. This procedure, intuitively, makes good sense, provided only that it is clear that the contradictory conclusion of the subproof is indeed "absurd" and unacceptable.

Like conditional proof, reductio ad absurdum can occur either on its own, as above, or within a more complicated argument, where other premises are in play. Those other premises may well be used in the work of deriving the contradiction that is the key step of the reductio. The summary conclusion (here, $\sim p$) will deny the last premise taken, not any other premise or any other statement, and that premise (p) will be discharged.

Principles of Inference 37

The procedure of reductio ad absurdum provides us with a second strategy of proof. The first was: to prove a conditional statement, take its antecedent as premise and try to prove its consequent. The second is: to prove the denial of a statement, take the statement itself as premise and try to derive a contradiction.

The other rule for negation: Double Negation, or DN, is quite different in character:

> Double Negation: A statement and its double negation are interchangeable.

We will deal with this in three parts. In the course of this discussion a number of alternative rules for double negation (DN', RED', etc.) will be mentioned and discussed, but not adopted. The intention is to explain and justify the rule DN, as stated here.

First, we note that the double negation of any statement is deducible from that statement, as follows:

```
 *  1  p                     PREM
**  2  ~p                    PREM
**  3  p ∧ ~p                1, 2 ADJ
 *  4  ~~p                   2-3 RED
                             QED
```

and we can go on to prove a theorem, as follows:

```
    5  p → ~~p               1-4 CP
                             QED
```

For example,

```
 *  1  Let us suppose that it is raining.    PREM
**  2  Let us suppose, further, that it is
       not raining.                          PREM
**  3  In that case it is both raining
       and not raining.                      1, 2 ADJ
```

But that is impossible. So, given our first premise (that it is raining), our *last* premise (that it is not raining) must be mistaken:

* 4 It is not not raining.   2–3 RED

So, summarizing, we conclude that

5 If it is raining, it is not not raining.   1–4 CP

    QED

Accordingly, this "part" of the rule DN is redundant. The converse, however, is not so deducible and not redundant; it is, in fact, debatable. A long tradition in logic, going back at least as far as the discussion of the sea battle tomorrow in Aristotle's *De interpretatione*, and also in mathematics, concerned primarily with existence theorems, questions the legitimacy of deducing a statement from its double negation. Some twentieth-century mathematicians, known as "intuitionists", have developed a strong body of mathematics that does without it. I will not, however, deal with these questions here, and will assume henceforward that nonintuitionistic double negation is indeed OK. We could state this straightforwardly:

DN': Any statement may be derived from its own double negation.

Accordingly,

$$\frac{\sim\sim p}{\therefore\ p}\ \text{DN}'$$

Or we could, instead, introduce a rule for ~ elimination, coordinate with RED, as follows:

REDuctio ad absurdum': The process of deriving a direct contradiction, in a rule-bound way, from the denial of a statement, justifies the assertion of the statement in question.

Accordingly,

* ~p
* .
* .
* .
* $q \wedge \sim q$
∴ p   RED'

(From the same premise and pattern of argument RED would yield $\sim\sim p$.) DN' can be justified by RED' (or vice versa), and so, using RED along with either RED' or DN', we could show that a statement can be deduced from its double negation, and vice versa. We have chosen instead to introduce a stronger rule: p and $\sim\sim p$ are interchangeable. How does this work?

What has been said so far is that a statement and its double negation are interchangeable *as steps in argument:* given p, $\sim\sim p$ is deducible from it, and vice versa. But DN tells us that a statement and its double negation are interchangeable *wherever they occur*, in the contexts we are interested in; that is, p and $\sim\sim p$ are interchangeable also as component parts of the compound statements with which we shall be working. From: if it rains, it is not the case that we will not go to the movies, it follows that we will go to the movies if it rains.[6]

To drop the clumsy locution of double negation where it turns up in argument or to introduce it on occasion when it is convenient for our purposes seems, intuitively, to be reasonable procedure – in conversation, we often do not notice when we do it – but, in careful reasoning, it is important to recognize use of DN, and also to be aware of its justification. A statement and its double negation are interchangeable because they are equivalent, because they "come to the same thing". In special contexts, where there is question about whether they do "come to the same thing," a philosophical question may arise. Although we shall set aside such questions, as beyond the scope of this book, we shall nonetheless be careful to be explicit when we use DN and to be aware of its usefulness.

Before we turn to the principles for "or" let us establish two more theorems concerning negation. These are '$(p \wedge \sim p) \rightarrow q$',

---

[6] A "metalogical" proof can be provided to show that, if two statements are provable from each other, they are interchangeable not only as steps in argument but also as components of compound statements. DN is a special case of this general principle of the interchangeability of equivalents, which will be discussed later in this chapter. The metalogical argument that supports it is, however, beyond the scope of this book.

which tells us that a conditional with contradictory antecedent is valid, regardless of its consequent, and '$\sim p \to (p \to q)$', which tells us that a conditional is valid if its consequent is also a conditional and the two antecedents contradict each other.

To Prove: $(p \wedge \sim p) \to q$

*	1	$p \wedge \sim p$	PREM
**	2	$\sim q$	PREM
**	3	$p \wedge \sim p$	1 REPEAT
*	4	$\sim \sim q$	2–3 RED
*	5	q	4 DN
	6	$(p \wedge \sim p) \to q$	1–5 CP
			QED

To Prove: $\sim p \to (p \to q)$

*	1'	$\sim p$	PREM
**	2'	p	PREM
***	3'	$\sim q$	PREM
***	4'	$p \wedge \sim p$	2', 1' ADJ
**	5'	$\sim \sim q$	3'–4' RED
**	6'	q	5' DN
*	7'	$p \to q$	2'–6' CP
	8'	$\sim p \to (p \to q)$	1'–7' CP
			QED

Alternatively, in proving the second theorem, we could have used the first, continuing the deduction shown above, as follows:

	6	$(p \wedge \sim p) \to q$	(See above)
*	7	$\sim p$	PREM
**	8	p	PREM
**	9	$p \wedge \sim p$	8,7 ADJ
**	10	$(p \wedge \sim p) \to q$	6 REPEAT
**	11	q	10, 9 MP
*	12	$p \to q$	8–11 CP
	13	$\sim p \to (p \to q)$	7–12 CP
			QED

It may be useful to remind oneself that repeating step 6, which carries no stars, at step 10, which carries two, is quite legitimate, even though "repetition" the other way round would be wrong. At a future step 14 we could use step 6 if we had reason to, but not step 10.

Rules for "Or"

We come now to the rules for "or", which we write '∨' ('v' for "vel" in Latin). Unfortunately, "or" is ambiguous in English (and in many other languages) as between the "exclusive or" and the "nonexclusive or" – although it is much more often used in the nonexclusive sense. If 'or' is taken nonexclusively, '--- or ...' means that at least one of --- and ... is the case; if 'or' is taken exclusively, '--- or ...' means that precisely one of --- and ... is the case, not both of them. In order to avoid this ambiguity, we *stipulate* that 'or' will be used hereafter, at least within our formal arguments, in the weaker, or nonexclusive sense. (Note that 'or', as used three words before this parenthesis, has neither sense; the comma that precedes it marks it as the 'or' of apposition, which is short for "in other words": the weaker sense *is* the nonexclusive sense.)

'Or' may be ambiguous; '∨' is not. By '∨' we shall mean one or the other, with no indication, one way or the other, about both. Like conjunction, disjunction is symmetrical; it does not matter which component is mentioned first.

A *disjunction* is a compound statement built from other statements by means of the connective "or" ('∨'); a disjunction affirms that at least one of its *disjuncts*, or disjunctive components, holds; it provides no information as to which or how many.

To use a disjunction we investigate the consequences of every disjunctive component, taken separately, and try to discover whether these consequences coincide. If they do coincide we can use the principle of DILemma.

To establish a disjunction, on the other hand, we need to deal with one disjunct only: if any one disjunct has been supposed or proved, regardless of the others, we can derive the disjunction.

The principle of DILemma, or ∨ elimination, like CP and RED, involves the taking and discharging of premises. It differs from those rules in that two or more premises are discharged at the same time, and also in that the premises taken are *dictated* by the disjunctive statement with which the argument begins. The list of premises taken must match – and exhaust – the list of disjunctive components of that disjunction.

DILemma: Given a disjunction, if each of its disjuncts can be shown by a separate argument to lead to the same conclusion, that conclusion follows from the disjunction.

For example,

```
        *   1  p ∨ q           PREM
    L * 2  p              PREM
    L * 3  .
    L * 4  .
    L * 5  .
    L * 6  r

    R * 7  q              PREM
    R * 8  .
    R * 9  .
    R * 10 .
    R * 11 r

    ∴ * 12 r              1, 2–6, 7–11 DIL
```

or, perhaps more graphically,

```
            * 1  p ∨ q  PREM

    L * 2  p  PREM              R * 7  q  PREM
    L * 3  .                    R * 8  .
    L * 4  .                    R * 9  .
    L * 5  .                    R * 10 .
    L * 6  r                    R * 11 r

            ∴ * 12 r   1, 2–6, 7–11 DIL
```

The word "DILemma" – short for "pair of lemmas", for such subproofs as 2–6 and 7–11 are "lemmas" – could of course be replaced by "trilemma" if there are three disjuncts in the disjunction from which we start or by "tetralemma" if there are four. Indeed the rule as I have stated it would be more aptly labeled "polylemma" – "poly" for "many" in Greek. But we shall not bother with such locutions, and cite instead the traditional term "dilemma." Note that 'L' (for "left") and 'R' (for "right") here function like stars: they make clear which premises are in play (any temptation to use steps from one lemma in another must be resisted). In a trilemma a third letter, say 'M' for "middle", would be useful; any unambiguous bookkeeping system will do. At the point at which the rule is invoked, it must be clear that every disjunctive component has been investigated.

The principle of dilemma provides us with a third strategy of proof: To use a disjunction in a deduction, take each of its disjuncts as a premise separately and try to prove the same conclusion using each of them, one at a time, without regard to the others. Plan to discharge all these premises, together.

The second rule for "or" is the rule of THINning (from the German 'Verdünnung'[7]), or $\vee$ introduction:

THINning: A disjunction follows from any one of its disjunctive components.

Accordingly,

$$\frac{p}{\therefore p \vee q} \text{ THIN} \qquad \frac{q}{\therefore p \vee q} \text{ THIN}$$
$$\text{etc.}$$

It is clear, of course, that, given that it is raining, it *is* either raining or snowing (although this would *not* follow if "or" had been

---

[7] Gerhard Gentzen, "Untersuchungen über das logische Schliessen", *Mathematische Zeitung*, vol. 39 (1934/5): 176–210, page 192. This essay is translated and reprinted in M. E. Szabo, *The Collected Works of Gerhard Gentzen* (Amsterdam: North-Holland, 1969), pp. 68–131; see p. 77 for Gentzen'n rules of inference, pp. 31 and 83 for his use of the word 'thinning'.

taken in the exclusive sense: given that it is raining, it follows that it is raining or snowing, but hardly that it is not doing both). Since what '$p \vee q$' means is that at least one of p and q is true and since the premise tells us that one of them *is* true, use of the rule of thinning cannot lead from truth to falsehood.

It is perhaps less clear why the rule of thinning would be useful. The conclusion is in an obvious way less informative (thinner) than the premise. How would it be useful?

Let us take an example. Suppose we know that if it rains or snows, school will be closed. We see that it is raining, and we conclude without hesitation that school will be closed. How do we justify that conclusion? Modus ponens is not straightforwardly available to us, because we do not have a conditional "taken together with its antecedent"; the second premise, "It is raining," does *not* match the antecedent, which is "It is raining or snowing". But the antecedent can be derived from the second premise, by the rule of thinning. Formally,

$$*\begin{cases} 1 & (r \vee s) \to c \\ 2 & r \end{cases} \quad \text{PREM}$$
$$*\ 3 \quad r \vee s \qquad\qquad 2 \text{ THIN}$$
$$*\ 4 \quad c \qquad\qquad\qquad 1, 3 \text{ MP}$$
$$\qquad\qquad\qquad\qquad\quad \text{QED}$$

Another use of the rule of THINning is to "force" a conclusion in a dilemma argument where the lemmas do not all easily yield the "same" conclusion; we use THINning as a last step in each of the lemmas. For example, given the premises:

$$*\begin{cases} 1 & p \vee q \\ 2 & p \to q \\ 3 & q \to s \end{cases} \quad \text{PREM}$$

we construct the following deduction:

```
L *  4  p           PREM (for DIL)
L *  5  r           2, 4 MP
L *  6  r ∨ s       5 THIN
R *  7  q           PREM (for DIL)
R *  8  s           3, 7 MP
R *  9  r ∨ s       8 THIN
  * 10  r ∨ s       1, 4-6, 7-9 DIL
                    QED
```

Notice steps 6 and 9.

This completes the list of basic principles of inference for propositional logic, principles that make explicit the ways in which the four basic propositional connectives are properly used in deduction. As we have seen, these connectives are:

  →   if__then ...   which generates a conditional
  ∧   and   which generates a conjunction
  ~   not   which generates a negation
  ∨   or   which generates a disjunction

Beside the nine basic principles we shall need six supplementary rules, to which we now turn. Two of these, INTerchange and THEOREM, allow us to use the body of logical culture that we build up as we work. Four others are familiar short cuts, redundant in that their reliability can be proved by means of the basic rules – anything that can be done by means of them can be done without them – but extremely convenient. And we shall need a fifth connective, defined by means of the others and so in theory redundant, but important both practically and conceptually:

  ↔   if and only if   which generates a biconditional

The biconditional will be defined and discussed in connection with the rule of INTerchange.

The Rule THEOREM

THEOREM: Any instance of a theorem, a statement form that has been established on logical grounds alone, may be introduced as a step in proof at any point in a deduction.

We have already taken note of a number of theorems, among them $p \to p$, $p \to \sim\sim p$, $p \to (q \to p)$, $\sim p \to (p \to q)$, $(p \wedge \sim p) \to q$. Clearly, it would be silly to prove these over and over every time we wished to use them or to go to the trouble of proving their instances, say, $(r \vee s) \to (r \vee s)$ or $(a \wedge b) \to \sim\sim(a \wedge b)$. We save ourselves this unnecessary labor by reminding ourselves of the results of work already done; we cite a theorem as a *short cut* in deduction.

The rule THEOREM authorizes us to use any theorem, once established, and indicates how this is to be done. We introduce an instance of the theorem as a step in proof, either at the beginning of a deduction or wherever it would be convenient. If it is introduced at the beginning it will require no stars; at a later point it will carry whatever stars are appropriate to its position (no star is added).

By way of nomenclature, we should note that the theorems of propositional logic (Part I of this book) and their instances are ordinarily called "tautologies." A statement is "tautologous," or a "tautology" if it can be shown to be true, regardless of any facts, by the methods of propositional logic.

Let us pause for a moment to discuss what is meant by an "instance" of a theorem. We have been using the notion of instantiating, or deriving instances, informally, since the first pages of this chapter, primarily in connection with the use of the rules. '$p \to q$' is an *instance* of a conditional statement. So are 'If it rains I will wear my umbrella' and 'I will wear my umbrella if it either rains or snows' and '$(p \vee q) \to (q \vee p)$', etc.; all of these are also instances of the symbolic '$p \to q$'. Modus Ponens tells us that from *any* conditional statement, taken together with its antecedent, its consequent follows, that is, that q can be derived from $p \to q$, taken together with p, *whatever statements we put in place of p and q*, without regard to the subject matter or complexity of these statements and regardless of whether they are different or alike. Every principle of inference affirms that all the arguments that are its instances are valid. These principles of inference, in other words, make quite general claims.

Theorems likewise make quite general claims. Since every theorem has already been proved to be valid, a similar proof of validity can be provided for any instance of it. We substitute for the variables of the original formulation the relevant variables or statements or statement forms that occur in the instance and then carry out the proof exactly as before. This can always be done, provided that the instance has been built correctly, that is, provided that whatever the original theorem said about, say p or q, the purported instance says about the substitutes for p and q.

Careful uniform substitution is our guarantee that the instances will be built correctly. Wherever a variable is repeated in the original its substitute must be repeated in the instance, and everything else in the original and the instance must be the same – otherwise we will have "changed the subject", illicitly. For example, it would be a mistake to think that '$p \to (p \vee q)$' is an instance of '$p \to p$', even though both are theorems, because in '$p \to (p \vee q)$' the antecedent and the consequent are not alike. Any proof for '$p \to (p \vee q)$' will be quite different from the proof we have given for '$p \to p$'. On the other hand, '$(p \vee q) \to (p \vee q)$' is an instance of '$p \to p$'; '$p \vee q$' has been uniformly substituted for 'p' both in the consequent and in the antecedent. The proofs for the two theorems will be alike, as the reader can verify. In ordinary argumentation, whenever we use a rule or a general claim, we take instances intuitively and for the most part reliably. Nonetheless it is wise to remind ourselves that this process must be carried out with care. As we have seen, '$p \to p$' is an instance of '$p \to q$', but '$p \to q$' is not an instance of '$p \to p$'. '$(p \wedge q) \to r$' is an instance of '$(q \wedge p) \to s$', and vice versa. '$(p \wedge q) \to p$' is an instance of '$(p \wedge q) \to r$', but not vice versa. '$a \to (b \wedge c)$' is an instance of '$p \to q$', but not of '$p \wedge q$'. Etc. Reliability must be learned through practice.

Let us illustrate the use of tautologies in deduction by carrying out a deduction using the rule THEOREM. It will be clear to the reader that we could have done this deduction without it, but it would be comparatively inconvenient.

We shall be claiming, notably in chapter 2, that a conditional statement $(p \to q)$ is equivalent to the disjunction $(q \vee \sim p)$ of the consequent (q) and the negation of the antecedent $(\sim p)$. Accordingly, the following argument should be valid:

> Either Mary is in danger or she is not crossing the street against the light.
> Therefore, if Mary is crossing the street against the light, Mary is in danger.

$d \vee \sim c$ To Prove: $c \to d$

Our strategy of proof will be dilemma.

```
  * 1  d ∨ ~c                 PREM
L * 2  d                      PREM (for DIL)
L * 3  d → (c → d)            THEOREM [p → (q → p)]
L * 4  c → d                  3, 2 MP
R * 5  ~c                     PREM (for DIL)
R * 6  ~c → (c → d)           THEOREM [~p → (p → q)]
R * 7  c → d                  6, 5 MP
  * 8  c → d                  1, 2-4, 5-7 DILemma
                              QED
```

We can, if we choose, take one more step, establishing another theorem:

     9   $(d \vee \sim c) \to (c \to d)$ 1-8 CP

                         QED

This is one half of an equivalence which we shall be calling "IF".

This half is ordinarily written:

$$(\sim p \vee q) \to (p \to q)$$

[The equivalence is:

         IF: $(p \to q) \leftrightarrow (\sim p \vee q)$

where '$\leftrightarrow$' means "if and only if" (see below).]

The Law of the Excluded Middle

We next take note of one more theorem, which is of some historical interest and logical utility as well: the Law of the Excluded Middle, or *Tertium non datur*, $p \vee \sim p$. We deal first with its use, then with its deduction.

The following is a familiar pattern of argument:

If I go to class I will learn nothing.
If I do not go to class I will learn nothing.
Therefore, I will learn nothing.

$$\begin{cases} c \to n \\ \sim c \to n \end{cases}$$
$$\therefore n$$

This feels like a dilemma, but the disjunction needed for a dilemma is missing. We use the law of excluded middle to fill the gap:

```
         1  c ∨ ~c              THEOREM
   * { 2  c → n                 PREM
       3  ~c → n
   L * 4  c                     PREM (for DIL)
   L * 5  n                     2, 4 MP
   R * 6  ~c                    PREM (for DIL)
   R * 7  n                     3, 6 MP
     * 8  n                     1, 4–5, 6–7 DIL
                                QED
```

We could, alternatively, have written the first three lines:

```
   * { 1  c → n                 PREM
       2  ~c → n
     * 3  c ∨ ~c                THEOREM
```

The basics of the proof would be the same. The point is that '$c \vee \sim c$', being an instance of a theorem and not a supposition, requires no star in the first case. In the second case it carries the star appropriate to its position; its introduction as a step in proof involves neither the taking nor the discharging of any premise.

We need now to establish the law of the excluded middle:

To Prove: $p \vee \sim p$

We begin by trying to map out a strategy. We cannot use Conditional Proof, because that strategy would yield a conditional, and '$p \vee \sim p$' is not a conditional. If we had reason to believe that p, we could derive $p \vee \sim p$ by THINning, but we have no reason to believe that p. Likewise, if we had reason to believe that $\sim p$, we could derive '$p \vee \sim p$' by THINning, but we have no reason to believe $\sim p$ either. If we had another disjunction to work from, we might be able to derive $p \vee \sim p$ by DILemma, but we have no such disjunction. We recall an old mathematician's injunction: "When in doubt, try reductio." We try reductio. So our first step will be to take as premise the denial of what we wish to establish:

```
*  1  ~(p ∨ ~p)           PREM (Neither p nor not p.)
```

We try to think through the argument. Assuming that it is neither raining nor not raining, I ought to be able to show that it is not raining ($\sim p$), for that is part of what the premise seems to say. Perhaps I can show *that* by reductio. And, if I can accomplish that, perhaps I can show, similarly, that it is not not raining ($\sim \sim p$). But then I will be contradicting myself (claiming both that it is not raining and that it is not not raining), which is what I wanted. (My strategy, in each of two subproofs, will also be reductio.) Formally, it looks like this:

To Prove: $p \vee \sim p$

```
*    1  ~(p ∨ ~p)                    PREM
**   2  p                            PREM
**   3  p ∨ ~p                       2 THIN
**   4  (p ∨ ~p) ∧ ~(p ∨ ~p)         3, 1 ADJ
*    5  ~p                           2-4 RED
                                     (first achievement)
**   6  ~p                           PREM
**   7  p ∨ ~p                       6 THIN
**   8  (p ∨ ~p) ∧ ~(p ∨ ~p)         7, 1 ADJ
*    9  ~~p                          6-8 RED
                                     (second achievement)
```

*	10	$\sim p \land \sim\sim p$	5, 9 ADJ
	11	$\sim\sim(p \lor \sim p)$	1–10 RED
	12	$p \lor \sim p$	11 DN
			QED

(This may look to the reader like a dilemma; it is not.) We note that, to prove the law of excluded middle, we required, in an essential way, use of the principle of double negation (step 12), in particular the controversial move of dropping a double negative. We note also that this proof is clumsy and repetitious. We can make it a little more elegant in either of two ways:

(1) *	1	$\sim(p \lor \sim p)$	PREM
**	2	p	PREM
**	3	$p \lor \sim p$	2 THIN
**	4	$(p \lor \sim p) \land \sim(p \lor \sim p)$	3, 1 ADJ
*	5	$\sim p$	2–4 RED
**	6	$\sim p$	PREM

Similarly,

*	9	$\sim\sim p$	6–8 RED
*	10	$\sim p \land \sim\sim p$	5, 9 ADJ
	11	$p \lor \sim p$	1–10 RED, DN
			QED

(2) *	1	$\sim(p \lor \sim p)$	PREM
**	2	p	PREM
**	3	$p \lor \sim p$	2 THIN
**	4	$(p \lor \sim p) \land \sim(p \lor \sim p)$	3, 1 ADJ
*	5	$\sim p$	2–4 RED
*	6	$p \lor \sim p$	5 THIN
*	7	$(p \lor \sim p) \land \sim(p \lor \sim p)$	6, 1 ADJ
	8	$p \lor \sim p$	1–7 RED, DN
			QED

So the law of the excluded middle, which is often taken to be axiomatic, is provable without axioms, by the principles of inference we are using (which include DN).

The Rule of INTerchange

INTerchange: Any two statements that have been shown to be equivalent, or deducible one from the other, are interchangeable.

We define a *biconditional* statement, a statement of the form 'p if and only if q' (Mary will go to the party if and only if John invites her; The water in that pot will boil if and only if its temperature is 212 degrees Fahrenheit; Jane sleeps until 10:00 A.M. if and only if today is Saturday) as the conjunction of two conditionals, one in each direction:

$$(p \leftrightarrow q) =_{\text{def}} [(p \to q) \wedge (q \to p)] \qquad \text{DEF} \leftrightarrow$$

And we define *equivalence* as *validity* of the biconditional. If two statements can validly, reliably, be derived one from the other, they effectively "make the same claim" and so can be interchanged, in any of the contexts in which we are interested, just as a statement may be interchanged with its own double negation.[8]

The principle of INTerchange allows us to *use* an equivalence. For example, given that $p \vee q$ is equivalent to $q \vee p$, as we have seen in discussing the meaning of '$\vee$', the following little argument is valid:

If it rains or snows, school will be closed.
It is snowing or raining.
Therefore, school will be closed.

$*\begin{cases} 1 & (r \vee s) \to c \\ 2 & s \vee r \end{cases}$ PREM

* 3 $r \vee s$ 2 INT (Commutation)
* 4 c 1, 3 MP
                         QED

or

* 3' $(s \vee r) \to c$ 1 INT (Commutation)
* 4' c 3', 2 MP
                            QED

---

[8] See footnote 6, page 39.

Before an equivalence can be used, however, it must be established. A deduction for an equivalence consists of two conditional proofs, one in each direction. Here are two examples.

To Prove: $(p \vee q) \leftrightarrow (q \vee p)$ (Commutativity of $\vee$)

```
      *   1  p ∨ q                     PREM (left to right)
  L   *   2  p                         PREM (for DIL)
  L   *   3  q ∨ p                     2 THIN
  R   *   4  q                         PREM (for DIL)
  R   *   5  q ∨ p                     4 THIN
      *   6  q ∨ p                     1, 2–3, 4–5 DIL
          7  (p ∨ q) → (q ∨ p)         1–6 CP
                                       (first achievement)
      *   8  q ∨ p                     PREM (right to left)
```

Similarly,

```
          14  (q ∨ p) → (p ∨ q)        8–13 CP
                                       (second achievement)
          15  [(p ∨ q) → (q ∨ p)]
                ∧ [(q ∨ p) → (p ∨ q)]  7, 14 ADJ
          16  (p ∨ q) ↔ (q ∨ p)        15 DEF ↔
                                       QED
```

To Prove: $[p \to (p \wedge q)] \leftrightarrow (p \to q)$ (Absorption)

```
     *   1   p → (p ∧ q)               PREM (left to right)
    **   2   p                         PREM
    **   3   p ∧ q                     1, 2 MP
    **   4   q                         3 SIMP
     *   5   p → q                     2–4 CP
         6   [p → (p ∧ q)] → (p → q)   1–5 CP
                                       (first achievement)
     *   7   p → q                     PREM (right to left)
    **   8   p                         PREM
    **   9   q                         7, 8 MP
    **  10   p ∧ q                     8, 9 ADJ
     *  11   p → (p ∧ q)               8–10 CP
        12   (p → q) → [p → (p ∧ q)]   7–11 CP
                                       (second achievement)
```

13 $\{[p \rightarrow (p \wedge q)] \rightarrow (p \rightarrow q)\}$
　$\wedge \{(p \rightarrow q) \rightarrow [p \rightarrow (p \wedge q)]\}$　6, 12 ADJ
14 $[p \rightarrow (p \wedge q)] \leftrightarrow (p \rightarrow q)$　　13 DEF $\leftrightarrow$
　　　　　　　　　　　　　　　QED

Steps 13 and 14 can be taken together:

13′ $[p \rightarrow (p \wedge q)] \leftrightarrow (p \rightarrow q)$　　6, 12 ADJ, DEF $\leftrightarrow$
　　　　　　　　　　　　　　　QED

The two conditional proofs that constitute a proof of equivalence are occasionally similar, as in the first case shown above. More often they are quite dissimilar. In the Reminders for this chapter I will provide a list of equivalences, available for use in interchange, some proved in the text, some left as exercises for the reader.

It should be recognized that the rule INT, like DN, differs from the other rules of inference in that it is directionless: if step 7 of a deduction follows from step 3 of that deduction by INT (a repetition of) step 3 will follow likewise from step 7. And equivalences differ from other theorems in the same way: reversing what appears to the left and right of the main connective '$\leftrightarrow$' in a valid equivalence does not change that equivalence. A statement derived from another by INT does not just follow from the first; it says the same thing.

Some of these equivalences – commutativity, contraposition, de Morgan's laws, for example – are often referred to in the literature as "rules," to be distinguished from such rules of inference as SIMPlification or THINning only in that they are directionless. I shall keep such equivalences quite separate from the rules of inference, and I shall treat them not as rules of procedure, but as statements, as theorems, to be established by deduction (or by other methods) and stored for use as needed – or else used as needed and established if necessary on the side.

The rule of inference to be invoked in using these equivalences is INTerchange.

Four Auxiliary Rules

CHAIN: From a pair (or sequence) of conditional statements in which the consequent of one conditional is the antecedent of the next, the conditional statement that has the first antecedent as its antecedent and the last consequent as its consequent, follows as a conclusion.

Accordingly,

$$\begin{array}{l} p \to q \\ \underline{q \to r} \\ \therefore p \to r \end{array} \quad \text{CHAIN} \qquad \begin{array}{l} p \to q \\ q \to r \\ \underline{r \to s} \\ \therefore p \to s \end{array} \quad \text{CHAIN} \quad \text{etc.}$$

The correctness of this procedure has been established above, in the course of our explanation of the principle of conditional proof.

A successful deduction for a *rule* starts from sample premises, taken together and marked with *one* star, and ends with a typical conclusion, again marked with one star. Clearly, a similar deduction could be carried out in similar cases. Once we have done one such deduction we have learned that it can be done and how it can be done, and we do not need to do it over again.

In establishing an auxiliary rule, all that should be used are the basic rules of inference.

Modus Tollens: From a conditional statement taken together with the denial of its consequent the denial of the antecedent follows.

Accordingly,

$$\begin{array}{l} p \to q \\ \underline{\sim q} \\ \therefore \sim p \end{array} \quad \text{MT}$$

NonCONJunction: From the denial of a conjunction, taken together with a component of the conjunction, the denial of the other conjunct (or of the rest of the conjunction) follows.

Accordingly,

$$\frac{\sim(p \wedge q)}{\therefore \quad \sim q} \quad \text{NCONJ} \qquad \frac{\sim(p \wedge q \wedge r)}{\therefore \quad \sim(p \wedge r)} \quad \text{NCONJ} \quad \text{etc.}$$

ELIMination: From a disjunction, taken together with the denial of a disjunctive component, the other disjunct (or the rest of the disjunction) follows.

Accordingly,

$$\frac{p \vee q}{\therefore \quad q} \quad \text{ELIM} \qquad \frac{p \vee q \vee r}{\therefore \quad p \vee q} \quad \text{ELIM etc.}$$

Reminders for Chapter 1

This page and the following pages contain a list of the principles of inference; some memos on the star system, on honest reporting when premises are discharged, and on the use of parentheses; a list of theorems, and a list of equivalences available for use in deduction. These reminders should prove useful in doing the problems.

Rules of Inference, Chapter 1

Modus Ponens: From a conditional statement taken together with its antecedent the consequent follows as a conclusion.

Sample:
$$\frac{p \to q}{\therefore \quad q} \quad \text{MP}$$

Conditional Proof: From a correct derivation of a conclusion from a premise (or set of premises) the corresponding conditional: if the premise(s) then the conclusion, follows as a summary conclusion.

Samples:
$$* \ p$$
$$\vdots$$
$$\underline{* \ q \to q}$$
$$\therefore \ p \to q \quad \text{CP}$$

$$* \left\{ \begin{array}{c} p \\ q \end{array} \right.$$
$$\vdots$$
$$\underline{* \ r}$$
$$\therefore \ (p \land q) \to r \quad \text{CP}$$

SIMPlification: From a conjunction any of its conjunctive components follows.

Samples:
$$\dfrac{p \land q}{\therefore \ p} \ \text{SIMP} \quad \dfrac{p \land q}{\therefore \ q} \ \text{SIMP} \quad \dfrac{p \land q \land r}{\therefore \ q} \ \text{SIMP}$$

ADJunction: From any statements in play at any point in a deduction their conjunction follows.

Samples:
$$\begin{array}{c} p \\ \underline{q} \\ \therefore \ p \land q \end{array} \ \text{ADJ} \quad \begin{array}{c} p \\ q \\ \underline{r} \\ \therefore \ r \land p \land q \end{array} \ \text{ADJ}$$

REDuctio ad absurdum: From a correct derivation of a direct contradiction from a premise, the denial of that premise follows as a summary conclusion.

Sample:
$$* \ p$$
$$\vdots$$
$$\underline{* \ q \land \sim q}$$
$$\therefore \ \sim p \quad \text{RED}$$

Double Negation (DN): A statement and its double negation are interchangeable, whether as steps in a deduction or as components of such steps.

DILemma: Given a disjunction, if each of its disjuncts has been shown by a separate deduction (lemma) to lead to the same conclusion, that conclusion follows from the disjunction.

$$* \ 1 \ p \vee q \ \text{PREM}$$

L * 2   p   PREM   R * 7   q   PREM
L * 3   .   R * 8   .
L * 4   .   R * 9   .
L * 5   .   R * 10 .
L * 6   r   R * 11 r

∴ * 12 r   1, 2–6, 7–11 DIL

THINning: From any one (or more) of the disjunctive components of a disjunction the disjunction itself follows.

Samples: $\dfrac{p}{\therefore \ p \vee q}$ THIN $\dfrac{q}{\therefore \ p \vee q}$ THIN $\dfrac{p \vee q}{\therefore \ p \vee r \vee q}$ THIN

REPEAT: In the course of a deduction it is always legitimate to REPEAT as a step in proof any premise that is still in play or any step in proof whose premise is still in play.

This completes the list of the nine basic principles of inference.

THEOREM: Any instance of a theorem, once established, may be introduced as a step in proof at any point in a deduction.

The theorem used should be cited, either by example or by name.

INTerchange: Any pair of statements that have been shown to be equivalent, or deducible one from the other, may be interchanged, whether as steps in deduction or as components of such steps.

The equivalence used should be cited, either by example or by name.

A list of four auxiliary principles of inference appears on the last pages of the chapter and so need not be repeated here; they are CHAIN, MT, NCONJ, and ELIM.

The Standard of Truth in Packaging

(in analogy to standards of truth in packaging
enforced in pharmacies and grocery stores)

When the principle of Conditional Proof is used, the summary conclusion drawn is a conditional statement whose antecedent is (in fact) the last premise taken and whose consequent is (in fact) a conclusion drawn from that premise in the context of the deduction. The justification for the claim that this conditional holds in that context is the subproof that starts from that premise and yields that conclusion; the subproof must be reported, truthfully, in the citation to the right of the step in question. The subproof is summarized and set aside; its premise is discharged.

This requirement of accurate reporting holds likewise when premises are discharged by reductio ad absurdum or by dilemma. The temptation to misreport is particularly strong in reductio proofs and the results of misreporting particularly misleading. A reductio proof shows that the premise from which it starts is false; it does not guarantee the falsity of any other step or any other premise.

On the Star System

It is important in any argument to be quite clear in one's mind about what suppositions or assumptions one is making, that is, about what premises are in play at any point. The star system we have been using is intended to facilitate that kind of bookkeeping.

A star marks the introduction of a supposition, or premise, whether at the outset of a deduction or later in the course of the work, and is repeated at every step in the deduction until that premise is discharged (or, in the special case of a dilemma, set aside); it is dropped when the premise is discharged. As long as the star associated with a premise appears at a step in proof, that premise is said to be *in play*.

When two or more premises are taken together, marked by a brace ({) and one star (ordinarily at the beginning of a deduction), the premises are regarded as taken *in conjunction*. Accordingly, if they are discharged, they must be discharged together: the resulting summary conclusion, or corresponding conditional, will have as antecedent the conjunction of those premises.

In a dilemma (or trilemma, etc.) the letter (L, R, M, etc.) associated with a premise (a component of the disjunction) is treated *as a star*, similarly indicating that its associated premise is in play at every step at which the star-letter occurs and similarly persisting within its lemma. But, when the desired joint conclusion is reached, the star-letter is set aside (rather than dropped altogether) in favor of another letter as another disjunct is taken as a premise. These letters are designed to help the arguer (and his reader) keep the lemmas separate and to discourage (illegitimate) "crossover" between them. Only when every disjunct of the original disjunction has been used as a premise and each has yielded the desired conclusion can all these auxiliary premises be discharged together and all the star-letters dropped.

A deduction ending with one star at its conclusion reports that the conclusion follows from the premise(s) from which the deduction began. Such a deduction is successful. The notation "QED" is, accordingly, appropriate.

A successful deduction can always be continued, if desired, so as to yield, by conditional proof, a conditional statement with no stars, i.e., a theorem. The lack of any stars at a step of a correct deduction marks that step as a theorem (a theorem of propositional logic is a *tautology*).

A deduction that ends with two or more stars reports that its last step follows from whatever premises those stars are associated with, taken together; those premises are all still in play, and the deduction is incomplete or unsuccessful. The notation "QED" would, accordingly, be *in*appropriate.

Since the star system registers the taking and discharging of premises and so makes visually clear which premises are in play and which are not, the *fallacy of illicit repetition* can be avoided by careful attention to the stars. A step in proof may not be repeated or otherwise used – for modus ponens, adjunction, etc. – unless it (if it is a premise) or its premise is still in play, that is, unless the step of use carries at least the stars carried by the step cited.

On Parentheses

Parentheses are used for grouping; they tell us to think of what is between them as a package, as a whole. They are needed to dispel ambiguity. They are not needed when there is no ambiguity. Because '$(p \wedge q) \wedge r$' is equivalent to '$p \wedge (q \wedge r)$' and '$(p \vee q) \vee r$' is likewise equivalent to '$p \vee (q \vee r)$', parens in those cases are not needed; they can be dropped in favor of '$p \wedge q \wedge r$' and '$p \vee q \vee r$', respectively. Parens are not needed in strings of letters connected uniformly by '$\wedge$' (and) or '$\vee$' (or).

Parentheses are needed, however, whenever the arrow for "if" is repeated. '$p \to q \to r$' is quite meaningless, since $(p \to q) \to r$ and $p \to (q \to r)$ are significantly different. And '$p \to q \to r$' does *not* mean "if p then q and if q then r " on the misleading analogy of '$x < y < z$' in algebra or '$2 < 4 < 9$' in arithmetic. '$p \to q \to r$' is just bad grammar.

Parens are also needed whenever propositional connectives are mixed. "We will go to the movies if it rains, and we will have dinner together" is different in meaning from "If it rains we will go to the movies and have dinner together." In ordinary language this difference is sometimes conveyed by a difference in word order, sometimes by "telescoping" or by some other device; in symbols it is conveyed by a difference in parentheses:

$$(r \to m) \wedge d \text{ versus } r \to (m \wedge d)$$

The ambiguous '$r \to m \wedge d$', like '$p \to q \to r$', is always disallowed as ill-formed, ungrammatical. Likewise, '$p \wedge q \vee r$'.

Without parens the "not" sign '$\sim$' applies only to the single letter that follows it; with parens it applies to the whole grouping (compound sentence) that immediately follows it. Since '$\sim \sim p$' means '$\sim(\sim p)$' unambiguously, those parens are not needed.

Otherwise, in formalizing compound sentences, parentheses to indicate grouping are always needed. Unnecessary parens are occasionally a nuisance, but not a mistake. Missing parens that lead to ambiguity do constitute a mistake.

I have used square brackets throughout to cover statements that already contain parentheses, and occasionally I have used curved braces to cover statements that already contain brackets. This practice makes for ease in reading, but it is not a matter of principle.

On Presenting Deductions

In presenting a deduction it is helpful to make clear at the outset what it is that one is attempting to demonstrate.

Given: such-and-such (premises, if there are any)
To Prove: so-and-so (the desired conclusion)

If the problem is stated in ordinary language a first step is to define one's abbreviations, introducing single letters of the alphabet in place of *sentences* (e.g., let $r =$ it is raining) and being careful not to use the same letter more than once (along with 'r' for "it is raining," 'r' for "Ronald is wearing his raincoat" would not do.) It is then possible to exhibit the structure of the argument (or theorem), replacing the sometimes vague or long-winded locutions of ordinary language with the unambiguous symbolic connectives of sentential logic.

A deduction starts with a premise or with two or three premises taken together, marked with one star (at the left), or else with a theorem, without stars. The steps of the deduction are numbered sequentially. To the right of each step we cite the *justification*, or warrant, for that step.

Used as justifications, the principles THEOREM and PREMise require no further citation. Occasionally, however, it is helpful to mention a theorem by name or to spell it out. For example:

    1 $(s \to w) \vee \sim(s \to w)$ THEOREM
                                            (Excluded Middle)

or

* 7 $(r \vee s) \to [a \to (r \vee s)]$ THEOREM
                                            $[p \to (q \to p)]$

In taking a premise, it is helpful to indicate the purpose of taking that particular premise: "To Prove" such and such, or "For DILemma" or "For REDuctio."

If the justification for a step is neither PREM nor THEOREM, an earlier step (or steps or subproof) must be cited. DN, INT, REPEAT, SIMP, and THIN use one step only; MP, MT,

NCONJ, and ELIM each use two; and ADJ and CHAIN each use two or more. The justification of each step should make quite clear where that step comes from and by what rule. For example:

```
*   3   w ∧ s
*   4   s                        3 SIMP
*   5   s ∨ r                    4 THIN
    .
    .
    .
*  10   ~(s ∧ g)
*  11   ~g                       10, 4 NCONJ
```

When INTerchange is used, one earlier step only is cited: the step within which the interchange occurs. The step cited and the step justified must match exactly, except for the interchange. It is helpful to cite the equivalence used, if possible by name. Although the equivalences used are theorems, the rule of INTerchange should not be confused with the rule THEOREM. INT requires that a previous step in proof be cited; THEOREM does not.

Whenever a step in proof is cited, one should be careful to check whether that step is in play (see the memo on the star system).

When DILemma is used, one must be very careful to keep the lemmas separate with no fallacious crossover. A sample dilemma is as follows:

```
              *  1  p ∨ q    PREM

L  *  2   p   PREM           R  *  7   q   PREM
L  *  3   .                  R  *  8   .
L  *  4   .                  R  *  9   .
L  *  5   .                  R  * 10   .
L  *  6   r                  R  * 11   r
         ∴  * 12  r     1, 2–6, 7–11 DIL
                             QED
```

The premises of the lemmas must match – and exhaust – the disjunctive components of the disjunction. No step in the L lemma may be cited in the R lemma, or vice versa. At the conclusion of the dilemma proof, the disjunction, along with the various lemmas, is cited. The conclusion carries the stars of the disjunction, reflecting the fact that the disjunction has not been discharged, whereas the premises of the lemmas have been discharged.

When a premise is discharged (whether by CP or DILemma or REDuctio) what is cited is not a step in proof, but a subproof, or lemma; this is made clear by use of a dash rather than a comma. '4, 9' for ADJ or MP means "steps 4 and 9"; '4–9' denotes the subproof that starts from step 4 *as premise* and ends with step 9.

The citation for a given step should contain no earlier steps from which the steps cited have been derived (that information is provided by the star system) or other steps suggestive of the steps cited (such as the disjunction that serves as basis for a dilemma) or steps that seem to motivate the step being taken. A common error in notation is as follows:

$*\begin{cases} 1 & (p \vee q) \to r \\ 2 & p \end{cases}$ PREM

To Prove: r

$*\ 3\ p \vee q$ 1, 2 THIN   (Incorrect)
$*\ 4\ r$ 1, 3 MP

Step 1 should *not* be cited at step 3 (the rest of this little deduction is of course correct).

Justification – ground, or reason to believe – should be clearly distinguished from motivation – purpose, or reason to do – and individual steps used in derivation should be clearly distinguished from general background, or context. Step 3 above, $p \vee q$, is derived by thinning from premise 2, p, alone; so the citation should be '2 THIN'. The *purpose* of taking step 3 is to be able to use premise 1 at step 4; premise 1 here "inspires" step 3, but it does *not* justify it.

The notation 'QED' ("quod erat demonstrandum" or

"mission accomplished") marks the end of a successful deduction. If the intention was to show the validity of an argument, the last step of the deduction should be the conclusion of that argument, with one star, signifying the dependence of the conclusion on the premises of the argument (which carried one star). If the intention was to establish a theorem, the theorem should appear without any stars, indicating that all premises have been discharged.

Theorems

Any properly constructed instance of a theorem may be introduced as a step in proof at any point in any deduction. A deduction in which all premises have been discharged and which ends, accordingly, with a conclusion carrying no stars, reports that its conclusion is a theorem. Examples of theorems are:

$p \to p$ (law of identity)
$p \to \sim\sim p$ (double negation)
$\sim\sim p \to p$ (double negation)
$\sim(p \wedge \sim p)$ (noncontradiction)
$p \to (q \to p)$
$\sim p \to (p \to q)$
$(p \wedge \sim p) \to q$
$(p \wedge q) \to q$
$p \to (p \vee q)$
$(p \to r) \to [(p \wedge q) \to r]$
$(p \to r) \to [p \to (q \vee r)]$
$p \vee \sim p$ (excluded middle)

Proofs for some of these theorems and some of the equivalences that follow have been provided in the text or assigned as problems for the reader. The reader may wish to compile and keep an auxiliary list of such theorems for her own use. In using a theorem it is well to keep its proof in mind; use of a theorem is a short cut, which could always be replaced by a deduction in context.

On Equivalence and the Biconditional

A *biconditional*, a statement of the form:

so and so if and only if such and such

or

$$p \leftrightarrow q$$

is defined as the conjunction of the two conditionals:

so and so if such and such

$$q \to p$$

so and so only if such and such

$$p \to q$$

Accordingly,

$$(p \leftrightarrow q) =_{\text{def}} [(p \to q) \wedge (q \to p)] \qquad \text{DEF} \leftrightarrow$$

Equivalence is defined as validity of the biconditional. Two statement forms – and, accordingly, two statements – are *equivalent* if it has been *established as a theorem* that the biconditional holds between them. To establish an equivalence by deduction two deductions are required, one in each direction, so as to establish two implications, which are then conjoined. Such an equivalence may be cited, if convenient, as a step in proof. For example,

3   $[(r \wedge s) \to w] \leftrightarrow [\sim(r \wedge s) \vee w]$ \qquad THEOREM (IF)

The principle of INTerchange authorizes the interchange of any two statements that have been shown to be equivalent, in any of the contexts we are interested in. For example (using the same equivalence), the following might serve as a step in proof:

* 4 $q \to (s \to t)$
* 5 $q \to (\sim s \vee t)$ \qquad 4 INT (IF)

The citation of IF in each case is helpful to the reader, but not necessary.

Equivalences for Use in INTerchange

Absorption:	$[p \rightarrow (p \wedge q)] \leftrightarrow (p \rightarrow q)$
Exportation:	$[(p \wedge q) \rightarrow r] \leftrightarrow [p \rightarrow (q \rightarrow r)]$
Contraposition:	$(p \rightarrow q) \leftrightarrow (\sim q \rightarrow \sim p)$
	$(p \rightarrow \sim q) \leftrightarrow (q \rightarrow \sim p)$
Commutativity:	$(p \wedge q) \leftrightarrow (q \wedge p)$
	$(p \vee q) \leftrightarrow (q \vee p)$
Associativity:	$[p \wedge (q \wedge r)] \leftrightarrow [(p \wedge q) \wedge r]$
	$[p \vee (q \vee r)] \leftrightarrow [(p \vee q) \vee r]$
Redundancy:	$(p \wedge p) \leftrightarrow p$
	$(p \vee p) \leftrightarrow p$
IF:	$(p \rightarrow q) \leftrightarrow (\sim p \vee q)$
IFF:	$(p \leftrightarrow q) \leftrightarrow (\sim p \leftrightarrow \sim q)$
	$(p \leftrightarrow q) \leftrightarrow [(p \wedge q) \vee (\sim p \wedge \sim q)]$
Distribution:	$[p \wedge (q \vee r)] \leftrightarrow [(p \wedge q) \vee (p \wedge r)]$
	$[p \vee (q \wedge r)] \leftrightarrow [(p \vee q) \wedge (p \vee r)]$
	$[p \rightarrow (q \wedge r)] \leftrightarrow [(p \rightarrow q) \wedge (p \rightarrow r)]$
	$[p \rightarrow (q \vee r)] \leftrightarrow [(p \rightarrow q) \vee (p \rightarrow r)]$
	$[(p \vee q) \rightarrow r] \leftrightarrow [(p \rightarrow r) \wedge (q \rightarrow r)]$
	$[(p \wedge q) \rightarrow r] \leftrightarrow [(p \rightarrow r) \vee (q \rightarrow r)]$

Problems for Chapter 1

I suggest that, after the first four principles of inference (for '→' and '∧', pages 14–35) have been studied, the following deductions be done:

in section I, problem 1;
in section II, problems 1–5.

Then, after the principles for negation have been introduced (pages 35–41), the following exercises should be done:

in section I, problems 2 and 3;
in section II, problems 6–9.

After the principles for "or" have been introduced (pages 41–45), the following exercises should be done:

in section II, problems 10–13;

Then, after the rules THEOREM (p. 45) and INTerchange (p. 52) have been introduced, the following exercises should be done:

in section I, problem 4;
in section II, problems 14–16.

Finally, before going on to chapter 2, part of section III should be done, at least the first example; i.e., the deduction for the equivalence called "Exportation." The rest of section III can wait until, say, one is working on chapter 3.

I. Here is a list of sample arguments illustrating the four auxiliary rules of inference given on the last pages of this chapter. In each case define your abbreviations and exhibit the structure of the argument. Then deduce the conclusion from the premises given, taken together. Use only the nine basic rules of inference.

1. CHAIN Rule

If it snows I'll wear galoshes.
I'll get blisters if I wear galoshes.
If I get blisters my feet will hurt.
Therefore, if it snows my feet will hurt.

2. Modus Tollens

> If yesterday was Sunday, today is Monday.
> Today is not Monday.
> Therefore, yesterday was not Sunday.

3. NonCONJunction

> It is not the case that we are in Paris and everyone here speaks Russian.
> Everyone here does speak Russian.
> Therefore, we are not in Paris.

4. ELIMination

> This book is about biology or it is about geology.
> This book is not about biology.
> Therefore, this book is about geology.

> (Hint: One way to work out this last deduction is to begin by using a theorem, appealing to the rule THEOREM as well as to the nine basic rules. Carry out the deduction and then deduce the theorem separately, on the side. This procedure will show you how you can rewrite the deduction *without* the rule THEOREM: within the longer deduction use the *method* by which you deduced the theorem, in place of the theorem itself. That theorem was, after all, only a short cut.)

II. Show by deduction that each of the following arguments is valid. Define your abbreviations explicitly.

1. If Alice goes both to the grocery store and to the butcher shop she will get everything she needs.
   Alice will go to the butcher shop if she goes to the grocery store.
   Therefore, if Alice goes to the grocery store, she will get everything she needs.

2. If John brings a salad, Henry will enjoy his dinner.
   Therefore, if Mary brings a casserole and John brings a salad, Henry will enjoy his dinner.

3. If Mary brings a casserole and John brings a salad, Henry will enjoy his dinner.
   Mary is planning to bring a casserole.
   Therefore, if John brings a salad Henry will enjoy his dinner.

4. Mary is bringing a casserole and if John brings a salad Henry will enjoy his dinner.
   Therefore, if John brings a salad Mary will bring a casserole and John will enjoy his dinner.

5. If Mary goes to the grocery store on her way home from work, we will have chocolate cake for supper.
   Mary will go to the grocery store if she is not delayed.
   She will not be delayed if the #5 bus is running.
   Therefore, if the #5 bus is running, we will have chocolate cake for supper.

6. Henry will go swimming if the weather is nice.
   Henry will not go swimming if he stays late at his office.
   Therefore, if the weather is nice, Henry will not stay late at his office.

7. If Cleon lies the gods punish him and he suffers.
   If Cleon lies men reward him and he does not suffer.
   Therefore, Cleon does not lie.

8. If Alice does not go both to the grocery store and to the butcher shop she cannot get everything she needs.
   If Alice goes to the grocery store she will not go to the butcher.
   Therefore, Alice will not get everything she needs.

9. If I do not think, I am thinking.
   Therefore, I am thinking.

   (Descartes argued, in support of this *premise*, that, if he claimed not to be thinking, his claim showed that he was thinking. Your deduction will show that the argument above is valid. Accordingly, if you question the certainty of its conclusion, you should also question its premise.)

10. If Mary brings a casserole or John brings a salad, Henry will enjoy his dinner.
    Therefore, if John brings a salad, Henry will enjoy his dinner.
11. If Cleon lies the gods punish him and he suffers.
    If Cleon tells the truth men punish him and he suffers.
    Cleon either lies or tells the truth.
    Therefore, Cleon suffers.
12. If Cleon lies the gods punish him and he suffers.
    If Cleon tells the truth men punish him and he suffers.
    Cleon either lies or tells the truth.
    Therefore, either the gods punish Cleon or men punish him.
13. Alice will go either to the grocery store or to the butcher, or else she will go to the movies with Mary.
    If Alice goes to the butcher she won't get milk.
    If she gets meat she won't go to the grocer.
    If she and Mary go to the movies, Alice will buy neither meat nor milk.
    Therefore, Alice won't get meat or she won't get milk.
14. If Mary is to declare a major either in French or in Comparative Literature, she must pass the French exam.
    Mary cannot pass the French exam.
    Therefore, she cannot declare a major in Comparative Literature.
15. We are going either to a baseball game or to the movies.
    If Alice gets her way, we will not go to the baseball game.
    If Caroline gets her way, we will not go to the movies.
    Therefore, either Alice or Caroline will not get what she wants.
16. In discussing the use of the law of the excluded middle, we used $c \vee \sim c$ to justify the following argument:
    If I go to class I will learn nothing.
    If I don't go to class I will learn nothing.
    Therefore, I will learn nothing.

$$c \to n$$
$$\sim c \to n$$
$$\therefore \quad n$$

Do a deduction for that argument *without* using the law of excluded middle. One strategy might be reductio.

III. Establish the following equivalences by deduction. This requires two deductions in each case – from left to right and from right to left – and, finally, adjunction and use of the definition of '↔'. See the discussion of the rule of INTerchange, at the end of this chapter, for examples of such deductions. Use only the nine basic rules of inference, and, if convenient, the rule of INTerchange, which will allow you to build on the work you yourself have done.

In each case, also provide an ordinary-language example. One such example for Exportation might be: "If I pay the fee and pass the test I will get my driver's license" is equivalent to "If I pay the fee then if I pass the test I will get my driver's license."

1. Exportation: $[(p \land q) \to r] \leftrightarrow [p \to (q \to r)]$
2. Contraposition: $(p \to q) \leftrightarrow (\sim q \to \sim p)$
3. IFF: $(p \leftrightarrow q) \leftrightarrow (\sim p \leftrightarrow \sim q)$

Three equivalences for Distribution:

4. $[p \land (q \lor r)] \leftrightarrow [(p \land q) \lor (p \land r)]$
5. $[p \to (q \land r)] \leftrightarrow [(p \to q) \land (p \to r)]$
6. $[(p \lor q) \to r] \leftrightarrow [(p \to r) \land (q \to r)]$

7. Given the first distribution rule (problem 4 above), the following equivalence follows immediately:

$$[p \land (q \lor \sim q)] \leftrightarrow [(p \land q) \lor (p \land \sim q)]$$

Explain how.

Finally, show separately that the left-hand formula: '$p \land (q \lor \sim q)$' is equivalent to 'p' and that the right-hand formula: '$(p \land q) \lor (p \land \sim q)$' also is equivalent to 'p'. It is OK to use a theorem.

Chapter 2
Truth Tables and Truth Trees

We have seen in the last chapter a set of rules of inference by which valid arguments can be constructed. If a conclusion can be derived, step by step, from a set of premises, in accordance with those rules, then the argument from those premises to that conclusion will have been shown to be valid. If, on the other hand, we are presented with an argument and *wonder* whether it is valid, these rules may not be all that helpful. If the argument "feels" valid and we are able easily to devise a derivation, well and good. Again, if the argument "feels" invalid and we are able easily to find a counterexample – a situation in which the premises would be true but the conclusion false – also, well and good: the rules will not have been helpful, but we will have established that the argument is invalid. Suppose, however, that we set out to design a derivation and fail; we learn nothing. Perhaps the argument is valid, but we have not been lucky or smart or persistent enough to show it; perhaps the argument is invalid and we have set ourselves an impossible task. We do not know.

Accordingly, when our project is not to demonstrate *that* an argument is valid, but rather to discover *whether* or not it is valid, we must take a different tack. We develop a procedure for *testing*. And we note at the outset that any such testing procedure must include at least a systematic procedure for discovering a counterexample if there is one.

We begin by listing and redescribing the truth-functional connectives we have been discussing: $\rightarrow$, $\wedge$, $\sim$, $\vee$, and $\leftrightarrow$, which are used to build compound statements from other statements. A *statement* is capable of truth or falsity. If a statement is true, its *truth value* is True; if it is false, its truth value is False.

A *compound* statement – or the connective used to build it – is *truth-functional* if the truth value of the compound is a function of the truth values of its component parts, if specifying whether each component statement is true or false is sufficient to determine without question whether the whole compound is true or false. This property – truth-functionality – is extremely useful. Since our objective is to argue reliably, that is, in such a way as always to preserve truth, never to derive from premises that are true any conclusion that is not true, it will be useful to be able to calculate the truth values of those premises and conclusions. And if we can use that calculation to find out whether, in all those situations in which the premises are true, the conclusions are true also, we shall have found out whether we have argued in a truth-preserving way. We shall have *tested* our arguments.

Let us see how this works.

Truth Tables

We begin with "if", or '→'. We recall that a conditional does not tell us one way or the other whether its antecedent is true; nor does it tell us, one way or the other, whether its consequent is true. It tells us only that, provided that the antecedent is true, the consequent is true also; so it rules out the possibility that the antecedent be true with the consequent false. No other possibility is ruled out. Perhaps the antecedent is false; in that case the conditional gives us no information as to whether the consequent is true: maybe yes, maybe no. Perhaps the consequent is true; *this* gives us no information as to whether the antecedent is true; maybe yes, maybe no.

We summarize these observations in the following table:

Initial possibilities for antecedent and consequent:		Situation possible, given that the conditional is true?
antecedent	consequent	
true	true	yes
true	false	no
false	true	yes
false	false	yes

Truth Tables and Truth Trees 75

Given that the conditional is true, it is possible that both antecedent and consequent are true (top row) or that the antecedent is false and the consequent true (third row) or that both antecedent and consequent are false (bottom row). More succinctly, either the consequent is true (regardless of the antecedent; top row or third row) or else the antecedent is false (regardless of the consequent; bottom two rows).

This is an interesting conclusion. To say that "if such and such, then so and so" is to say that "either not such and such or else so and so", no more than that.

We noted earlier (at the beginning of chapter 1, in discussing the principles of inference for "if") that we shall be setting aside, as irrelevant to our work, any causal or other connections between the antecedents and the consequents of the conditional statements we deal with. These connections ordinarily serve to justify a person's belief that a conditional is true, and, even when we try to disregard them, they continue to flavor the English-language expression of that conditional. Still, they are not part of its meaning, not part of what it is intended to convey.

Accordingly, it should come as no surprise to the reader that the logician's truth tables for "if" make evident the meagerness of the information given by a conditional statement: all it tells us is that either its consequent is true or its antecedent is false.

This observation is in conformity with the equivalence IF which is listed in chapter 1 among the equivalences available for use in interchange, but which is only half established there. Let us establish it here. We start (recapitulating) from right to left:

To Prove: $(p \to q) \leftrightarrow (\sim p \vee q)$

```
        *   1  ~p ∨ q                    PREM
   L    *   2  ~p                        PREM (for DIL)
   L    *   3  ~p → (p → q)              THEOREM
   L    *   4  p → q                     3, 2 MP
   R    *   5  q                         PREM (for DIL)
   R    *   6  q → (p → q)               THEOREM
   R    *   7  p → q                     6, 5 MP
        *   8  p → q                     1, 2-4, 5-7 DIL
            9  (~p ∨ q) → (p → q)        1-8 CP
```

We turn next to the implication from left to right:

```
  *   10  p → q                          PREM
 **   11  ~(~p ∨ q)                      PREM (for RED)
***   12  p                              PREM (for RED)
***   13  q                              10, 12 MP
***   14  ~p ∨ q                         13 THIN
***   15  (~p ∨ q) ∧ ~(~p ∨ q)           14, 11 ADJ
 **   16  ~p                             12–15 RED
 **   17  ~p ∨ q                         16 THIN
 **   18  (~p ∨ q) ∧ ~(~p ∨ q)           17, 11 ADJ
  *   19  ~~(~p ∨ q)                     11–18 RED
  *   20  ~p ∨ q                         19 DN
      21  (p → q) → (~p ∨ q)             10–20 CP
      22  (p → q) ↔ (~p ∨ q)             9, 21 ADJ, DEF ↔
                                         QED
```

(It is worth noticing that, in the reductio subproofs, step 12 is discharged at step 16 and step 11 at step 19. The order of work is important.)

We now reconstruct the table for "→" in more traditional truth-table form, and then go on to construct truth tables for the other four connectives.

IF:

p	q	p → q
T	T	T
T	0	0
0	T	T
0	0	T

where 'T' stands for "true" and '0' for "false",[1] and where the right-hand column tells us that "$p \to q$" leaves open the possibility that both p and q are true (top row) and also the possibilities that p is false along with q true (third row) and that p is false

---

[1] W. V. Quine, in *Methods of Logic* (New York: Holt, Rinehart and Winston, 1st ed. 1950, 2d 1959, 3d 1972; 4th ed. Cambridge, Mass.: Harvard University Press, 1982), hereafter, *Methods*, uses '⊥' for false; many other books use 'F'; I favor '0', for reliability of reading; '0' does not look like 'T'.

along with *q* false (bottom row), but rules out the possibility that *p* be true with *q* false (second row). Given that "$p \to q$" is true, the truth values of its components are *not* known – there are many possible combinations. On the other hand, $p \to q$ will count as true in all cases except one: $p = T$ and $q = 0$ (second row), where it is false; so the conditional turns out to be truth-functional: if the truth values of the antecedent and consequent are given, we can look up the truth value of the compound '$p \to q$'.

Note also that, if a conditional is known to be *false*, only one possibility is open: that its antecedent is true and its consequent false (second row). This again reflects the fact that we are dealing with "bare" conditionals. To say that a conditional is false is not to deny some causal or other connection between its antecedent and its consequent, leaving open various possibilities; it is to claim that the one possibility ruled out by the conditional is indeed the case: its antecedent is true, and its consequent is false.

The truth table for "and", or '$\wedge$', is more straightforward. A conjunction – a compound built with "and" as its main connective – to be true, requires that both (or all) of its conjunctive components be true; the conjunction is false otherwise.

AND:

p	*q*	$p \wedge q$
T	T	T
T	0	0
0	T	0
0	0	0

Note that, given the *truth* of $p \wedge q$, the truth values of *p* and of *q* are determinate, but, given the *falsity* of $p \wedge q$, they are not. In the case of falsity, there are three possibilities: both components may be false (bottom row) or one or the other may be false and the other one true. Nonetheless, "and" is clearly truth-functional: if the truth values of the conjunctive components are given, we can look up or figure out the truth value of the compound.

The truth table for "not" is again straightforward: if a statement is true, its denial is false; if a statement is false, its denial is true. Note that acceptance of this claim involves acceptance of the principle of double negation, DN, including its controversial component: $\sim\sim p \to p$.

NOT:
p	$\sim p$
T	0
0	T

Since, as this table makes clear, wherever a statement is true, its denial is false, and wherever a statement is false, its denial is true, we can begin to construct truth tables for the *negations* of compound statements. First, double negation:

p	$\sim p$	$\sim\sim p$
T	0	T
0	T	0

and, if we wish, triple negation, etc.:

p	$\sim p$	$\sim\sim p$	$\sim\sim\sim p$
T	0	T	0
0	T	0	T

We remind ourselves that 'not not ...' is *not* emphatic for 'not'.

Next we construct the truth table for the denial of conjunction – in computer terminology, NAND:

NAND:

p	q	$p \wedge q$	$\sim(p \wedge q)$
T	T	T	0
T	0	0	T
0	T	0	T
0	0	0	T

Likewise, we construct the truth table for the denial of the conditional, which, we have already seen, will turn out to match the truth table for the conjunction of its antecedent with the denial of its consequent, the only case in which the conditional is false. Here is the truth table for NIF:[2]

---

[2] Herbert Bohnert, in *Logic: Its Use and Methods* (Washington, D.C.: University Press of America, 1977), and in his teaching, introduced the abbreviation 'NIF' in analogy with 'NOR' and 'NAND'.

NIF:

p	q	$p \to q$	$\sim(p \to q)$
T	T	T	0
T	0	0	T
0	T	T	0
0	0	T	0

In order to emphasize the "match" noted above, we construct the truth table for $p \wedge \sim q$:

p	q	$\sim q$	$p \wedge \sim q$
T	T	0	0
T	0	T	T
0	T	0	0
0	0	T	0

If we consolidate these two truth tables, the match between $\sim(p \to q)$ and $p \wedge \sim q$ becomes obvious:

p	q	$p \to q$	$\sim(p \to q)$	$\sim q$	$p \wedge \sim q$
T	T	T	0	0	0
T	0	0	T	T	T
0	T	T	0	0	0
0	0	T	0	T	0

The column for $\sim(p \to q)$ is built by "negating" the column for $p \to q$: putting 0 wherever $p \to q$ has T and T wherever $p \to q$ has 0. The column for $\sim q$ is built similarly by negating the reference column for q, and the column for $p \wedge \sim q$ is built by "conjoining" the columns for p and for $\sim q$: putting T wherever both p and $\sim q$ show T, and 0 everywhere else. Finally, we look at the two columns for $\sim(p \to q)$ and for $p \wedge \sim q$. The precise match is obvious. We have used truth tables to establish the equivalence:

NIF: $\sim(p \to q) \leftrightarrow (p \wedge \sim q)$

We could have established this equivalence by the methods of chapter 1. This is a lot easier.

We turn next to the truth table for '$\vee$', the main connective of disjunctive statements. As in chapter 1, disjunction will be construed nonexclusively: a disjunction is true if any of its disjunctive components is true, false otherwise.

OR:

p	q	$p \vee q$
T	T	T
T	0	T
0	T	T
0	0	0

This truth table is reminiscent of the truth table for the conditional, in that the "$p \vee q$" column contains three Ts and one 0. The truth table for the denial of a disjunction is likewise reminiscent of the truth table for the denial of a conditional:

NOR:

p	q	$p \vee q$	$\sim(p \vee q)$
T	T	T	0
T	0	T	0
0	T	T	0
0	0	0	T

The $\sim(p \vee q)$ column contains one 'T' only: where p and q are both false, that is, where $\sim p \wedge \sim q$ is true. Let us expand this truth table as we did that for $\sim(p \to q)$ and establish another equivalence:

p	q	$p \vee q$	$\sim(p \vee q)$	$\sim p$	$\sim q$	$\sim p \wedge \sim q$
T	T	T	0	0	0	0
T	0	T	0	0	T	0
0	T	T	0	T	0	0
0	0	0	T	T	T	T

The columns for $\sim(p \vee q)$ and for $\sim p \wedge \sim q$ again show a dead match; we have shown this equivalence:

NOR: $\sim(p \vee q) \leftrightarrow (\sim p \wedge \sim q)$

For completeness and orderliness of exposition, we go back to see whether a comparable equivalence can be established for NAND. The truth table for NAND reflects the fact that a conjunction is true only if all its conjuncts are true; it is false – and its denial true – if any of its conjuncts is false. So, what $\sim(p \wedge q)$ tells us is that either $\sim p$ or $\sim q$. We check this observation by building the expanded truth table for NAND:

p	q	$p \wedge q$	$\sim(p \wedge q)$	$\sim p$	$\sim q$	$\sim p \vee \sim q$
T	T	T	0	0	0	0
T	0	0	T	0	T	T
0	T	0	T	T	0	T
0	0	0	T	T	T	T

The columns for $\sim(p \wedge q)$ and $\sim p \vee \sim q$ are alike; so we have established the equivalence:

NAND: $\sim(p \wedge q) \leftrightarrow (\sim p \vee \sim q)$

This pair of equivalences: NOR and NAND, are traditionally known as "De Morgan's laws," after the nineteenth-century logician Augustus De Morgan, who drew attention to them.

We turn finally to "if and only if", or "iff", written '$\leftrightarrow$', the main connective of the *biconditional,* which was defined in chapter 1 as the conjunction of a conditional and its converse:

$$p \leftrightarrow q =_{\text{def}} [(p \rightarrow q) \wedge (q \rightarrow p)] \quad \text{DEF} \leftrightarrow$$

Let us pay attention for a moment to the locution 'if and only if'. 'Iff' is a philosopher's abbreviation; note that it is *not* emphatic for 'if' any more than '$\sim\sim$' is emphatic for '$\sim$'. '$p \leftrightarrow q$', 'p if and only if q', means, as one would expect, "*p* if *q*" *and* "*p* only if *q.*" Now it is easy to see that the right-hand conjunct above: $q \rightarrow p$, means "*p* if *q* ." But why does the left-hand conjunct, $p \rightarrow q$, tell us that "*p* only if *q*"? I think that the natural way to express "*p* only if *q*" in the symbols we have been using is to reinterpret it first as the conditional: "if not *q* then not *p*." If I tell you that we will go hiking only if the sun shines, you hear me as telling you that if the sun does not shine we will not go hiking. So "*p* only if *q*" could be symbolized as '$\sim q \rightarrow \sim p$'. But we saw in chapter 1 that $\sim q \rightarrow \sim p$, by Contraposition, is equivalent to and interchangeable with $p \rightarrow q$, and also that order in conjunction does not matter. So the biconditional $p \leftrightarrow q$, *p* if and only if *q*, *p* if *q* and *p* only if *q*, could be written as $(q \rightarrow p) \wedge (\sim q \rightarrow \sim p)$ or as $(q \rightarrow p) \wedge (p \rightarrow q)$ or as $(p \rightarrow q) \wedge (q \rightarrow p)$. Any of these formulations would do; the last one is traditional.

We saw, also at the end of chapter 1, that if a biconditional is shown to be *valid*, and so to constitute an equivalence, it can be used for interchange. [We have just been using the equivalence:

$$(p \rightarrow q) \leftrightarrow (\sim q \rightarrow \sim p) \quad \text{Contraposition}$$

in our exposition.] But not all true (or supposed) biconditionals are (or purport to be) valid; an "ordinary" biconditional (the water on the stove is boiling if and only if its temperature is 212 degrees Fahrenheit; I will go to the party if and only if I am invited) is more likely to be used in argument by way of the definition: one of its two constituent conditionals is derived from the biconditional by simplification and then used for modus ponens or chain reasoning or whatever.

It is important to be clear in every case about whether the claims we are prepared to make are conditional or biconditional; for a biconditional makes two conditional claims and a conditional makes only one. So $p \leftrightarrow q$ is not to be confused with either $p \rightarrow q$ or $q \rightarrow p$ (which, of course, are not to be confused with each other).

Let us now build up the truth table for the biconditional from its definition:

$$p \leftrightarrow q =_{\text{def}} [(p \rightarrow q) \wedge (q \rightarrow p)]$$

p	q	$p \rightarrow q$	$q \rightarrow p$	$(p \rightarrow q) \wedge (q \rightarrow p)$
T	T	T	T	T
T	0	0	T	0
0	T	T	0	0
0	0	T	T	T

Accordingly, we have

IFF:

p	q	$p \leftrightarrow q$
T	T	T
T	0	0
0	T	0
0	0	T

Supposing that p iff q, either both p and q are true (top row) or both p and q are false (bottom row); p and q are true together or false together. This observation is reflected in another of the equivalences noted in chapter 1:

IFF: $(p \leftrightarrow q) \leftrightarrow [(p \wedge q) \vee (\sim p \wedge \sim q)]$

Truth Tables and Truth Trees   83

The reader may wish to satisfy himself that this equivalence holds, either by deduction or by truth table.

Furthermore, if it is not the case that p iff q (second and third rows of the truth table for IFF), one is true, but not the other. So we have another truth table for negation:

NIFF:	p	q	$p \leftrightarrow q$	$\sim(p \leftrightarrow q)$
	T	T	T	0
	T	0	0	T
	0	T	0	T
	0	0	T	0

This, like the truth tables for NIF, NOR, and NAND, can be expanded to yield another equivalence (in this case three of them):

p	q	$p \leftrightarrow q$	$\sim(p \leftrightarrow q)$	$\sim p$	$\sim p \leftrightarrow q$	$\sim q$	$p \leftrightarrow \sim q$
T	T	T	0	0	0	0	0
T	0	0	T	0	T	T	T
0	T	0	T	T	T	0	T
0	0	T	0	T	0	T	0

p	q	$\sim p$	$\sim q$	$p \wedge \sim q$	$\sim p \wedge q$	$(p \wedge \sim q) \vee (\sim p \wedge q)$
T	T	0	0	0	0	0
T	0	0	T	T	0	T
0	T	T	0	0	T	T
0	0	T	T	0	0	0

Accordingly,

NIFF: $\sim(p \leftrightarrow q) \leftrightarrow (\sim p \leftrightarrow q)$
$\sim(p \leftrightarrow q) \leftrightarrow (p \leftrightarrow \sim q)$
$\sim(p \leftrightarrow q) \leftrightarrow [(p \wedge \sim q) \vee (\sim p \wedge q)]$

And we can show, furthermore, that

IFF: $(p \leftrightarrow q) \leftrightarrow (\sim p \leftrightarrow \sim q)$

A biconditional tells us that two statements are alike in truth value; their denials will, accordingly, also be alike in truth value. The denial of a biconditional tells us that two statements are different in truth value, and so the truth value of one is the truth value of the denial of the other.

Truth tables can be used directly to check the validity of truth-functional arguments. One builds up the truth table for the conjunction of the premises of the argument and the corresponding truth table for the conclusion. If there is any row where the premises come out true but the conclusion false, the argument is invalid – a counterexample has been found; if not, the argument is valid. For example, to test problem II-6 of chapter 1, p. 70, we build a truth table as follows:

1: Henry will go swimming (s) if the weather is nice (w).
2: Henry will not go swimming if he stays late at the office (o).
∴ If the weather is nice Henry will not stay late at the office.

			P1		P2	PREM		CON	OK?
s	w	o	$w \to s$	$\sim s$	$o \to \sim s$	P1 $\wedge$ P2	$\sim o$	$w \to \sim o$	
T	T	T	T	0	0	0	0	0	
T	T	0	T	0	T	T	T	T	yes
T	0	T	T	0	0	0	0	T	
T	0	0	T	0	T	T	T	T	yes
0	T	T	0	T	T	0	0	0	
0	T	0	0	T	T	0	T	T	
0	0	T	T	T	T	T	0	T	yes
0	0	0	T	T	T	T	T	T	yes

At every row at which both premises show "true" (rows 2, 4, 7, and 8) the conclusion shows "true" also.

In contrast, an invalid argument has at least one row at which the premises are all true, but the conclusion is false. For example, we may test an instance of the "fallacy of denying the antecedent", the fallacy corresponding to modus tollens, as follows.

Henry will go swimming if the weather is nice.
The weather will not be nice.
Therefore, Henry will not go swimming.

		P1	P2	PREM	CON	OK?
s	w	w → s	~w	P1 ∧ P2	~s	
T	T	T	0	0	0	
T	0	T	T	T	0	no
0	T	0	0	0	T	
0	0	T	T	T	T	

On the second row, the premises both show "true", but the conclusion is false. Accordingly, it is possible that Henry will go swimming if the weather is nice (P1) but that, although the weather is not nice (P2), he will nonetheless go swimming (the denial of the conclusion) – and so the argument is invalid.

This testing procedure works, but it can be extraordinarily cumbersome. Problem II-11 of chapter 1, for example, which is a classical dilemma, would require a truth table with 32 rows. One can devise short cuts. W.V. Quine has invented a much more elegant method, truth-value analysis,[3] based on the same definitions and, like truth tables, limited to propositional arguments. I shall present instead a variant of Richard Jeffrey's truth trees.[4] The method of truth trees can be extended smoothly to deal with quantificational arguments and has the additional advantage of appeal to the visual-minded.

---

[3] *Methods*, part I, chapter 5.
[4] Richard C. Jeffrey, *Formal Logic: Its Scope and Limits* (New York: McGraw-Hill Book Company, 1967), part I, ch. 4. The method recommended in the next few pages derives from Jeffrey's system, but is simplified for the benefit of readers with a minimum of mathematical sophistication. I have separated the rules for diagramming negations of compound statements from the four basic rules for diagramming conjunctions, disjunctions, conditionals, and biconditionals – this for the benefit of the visual-minded. In discussion with students over the years I have learned that much of the visual advantage of the tree method is lost if the rules for handling negation are introduced too soon.

Truth Trees

A *truth tree* is a diagram showing the various ways in which a statement or conjunction of statements can be true. A simple conjunction of two statements can be true in only one way: if both of the components are true. A conjunction is diagrammed up and down (like the trunk of a tree):

$$\text{AND:} \quad \begin{array}{|l} p \\ q \end{array}$$

The items placed on a tree are "atomic statements" – the simplest statements involved in a given context – or their negations. So $p \wedge \sim p$ is drawn:

$$\begin{array}{|l} p \\ \sim p \end{array}$$

This, of course, is not a real possibility; $p \wedge \sim p$ is self-contradictory; p and $\sim p$ cannot be true together. We register the impossibility with an **X**:

$$\begin{array}{|l} p \\ \sim p \\ \mathbf{X} \end{array}$$

This little tree is *closed*.

A simple disjunction of two statements can be true in either of two ways; $p \vee q$ is true if p is true, and it is true if q is true. The diagram *branches*:

OR: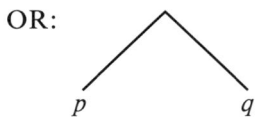

There is no contradiction here; both branches are *open*. Nor is there any contradiction in the diagram for $p \lor \sim p$:

Again, both branches are open.

The last three little diagrams have made visually evident that a statement of the form "$p \land \sim p$" is *inconsistent* (it cannot in any way be true; the only branch of its truth-tree diagram is closed), in contrast to such as "$p \lor q$" or "$p \lor \sim p$," which are *consistent* (it is logically possible that they be true; at least one branch of each truth-tree diagram is open).

Note that the word 'consistent', as here introduced and used (more or less consistently) in the literature of logic, is different in meaning from the word 'consistent' in most colloquial usage. A father is consistent if he praises or blames *all* his children for the same virtues or offenses. A statement is consistent if it can, in some circumstance, be true.

We shall be using truth trees primarily to see whether statements or conjunctions of statements are consistent, and secondarily – though more importantly – as an indirect method of checking the validity of both arguments and theorems, of discovering whether these arguments or theorems are valid.

Two more basic truth-tree diagrams will be needed: for "if" and for "if and only if." But, before we go on to consider those two basic diagrams, let us see how to use the first two for constructing slightly more complicated diagrams for combinations of "and" and "or". The diagram for $(p \land q) \land (r \land s)$ is

p
q

r
s

but neither order nor grouping is of any significance; so any of these would do:

88   Truth-functional Logic

$$\begin{vmatrix} q \\ p \\ r \\ s \end{vmatrix} \quad \begin{vmatrix} p \\ q \\ s \\ r \end{vmatrix} \quad \begin{vmatrix} r \\ s \\ p \\ q \end{vmatrix} \quad \begin{vmatrix} s \\ r \\ p \\ q \end{vmatrix}$$

The diagram for $(p \vee q) \vee (r \vee s)$ is:

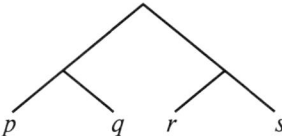

But again, neither order nor grouping matters; so either of these would do:

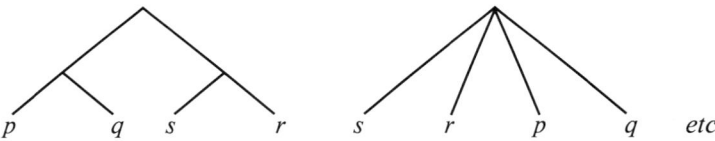

The first is easiest to follow, and we shall, accordingly, use this pattern unless considerations of space interfere. It should be remembered that the items entered on a truth tree are single letters or their negations.

When "or" and "and" are mixed, however, grouping does matter. The disjunction of conjunctions, $(p \wedge q) \vee (r \wedge s)$, will look like this:

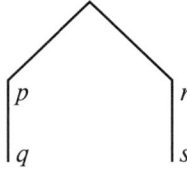

and the conjunction of disjunctions, $(p \vee r) \wedge (q \vee s)$, will look like this:

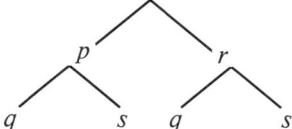

Note the difference in *pattern* between the last two diagrams: the first has two branches; the second has four.

Let us take an example. Let $p =$ Peter goes to the movies; $q =$ Quentin goes to the movies; $r =$ Rose goes to the movies; $s =$ Susan goes to the movies. Suppose we are informed that Peter and Quentin are going to the movies. In that case, these diagrams show us that both compounds are true; for p and q occur together on the left-hand branches of both trees. Remember that what a branch says is that the statements that appear on it constitute one way in which the statement being diagrammed comes out true. Peter and Quentin or Rose and Susan do go to the movies (first diagram), and Peter or Rose and Quentin or Susan do go to the movies as well (second diagram). But suppose instead that we are told that Quentin and Rose are going to the movies. In that case the second diagram tells us that the conjunction $(p \vee r) \wedge (q \vee s)$ is true: q and r occur together on the third branch of the tree. But, in the first diagram, neither branch contains both q and r; so we have been given no information as to whether $(p \wedge q) \vee (r \wedge s)$ is true.

The two compounds we are discussing have different "truth conditions", and this difference is reflected in the difference between their truth trees: in the two cases where the disjunction is true, the conjunction is true also, but not vice versa. (In both diagrams all branches are open; there is no negation that might lead to closure.)

Let us see now how closure affects the structure of a truth tree when it does occur. Let us diagram the "exclusive or", say, "Mary will have apple pie or cherry pie, but not both": $(a \vee c) \wedge \sim(a \wedge c)$. NAND tells us that this is equivalent to $(a \vee c) \wedge (\sim a \vee \sim c)$. This can be diagrammed in at least two ways:

Both trees show only two *open* branches: (*a* and ~*c*) and (*c* and ~*a*). If we wish to conjoin some further statement, say *p*, that statement will be "hung" from each of the open branches, and the closed branches will be disregarded:

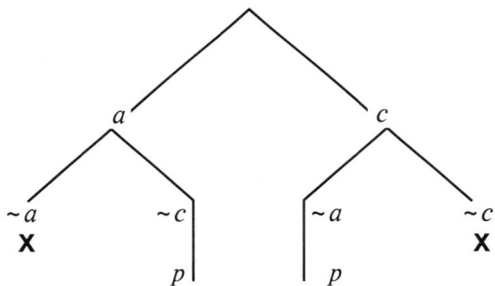

(Putting *p* on a closed branch would be wasted effort; the branch would still be closed; $a \wedge \sim a \wedge p$ is just as impossible as $a \wedge \sim a$.)

To set up the diagrams for the conditional and the biconditional, we now go back to their truth-table definitions. In building the truth table for the conditional, we saw that a conditional says that either its consequent is true or else its antecedent is false. This claim is expressed in the equivalence:

IF: $(p \rightarrow q) \leftrightarrow (\sim p \vee q)$

which is demonstrable, as we have seen, either by deduction or by truth table.

Accordingly, we have the diagram for $p \rightarrow q$:

IF: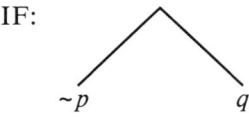

(Note that this can, if we prefer, be drawn:

but that

conveys something quite different; it is the diagram for $q \to p$.)

The truth table for the biconditional shows that either both components are true or both are false; so $p \leftrightarrow q$ is drawn:

We can check out this diagram for the biconditional by going back to its definition:

$$(p \leftrightarrow q) =_{\text{def}} [(p \to q) \wedge (q \to p)] \quad \text{DEF} \leftrightarrow$$

We can combine the diagram for $p \to q$:

with the diagram for $q \to p$:

to obtain the diagram for $(p \to q) \land (q \to p)$:

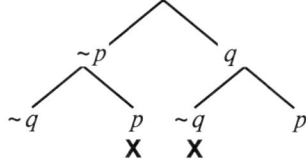

If we disregard the closed branches, what is left is a diagram with two open branches:

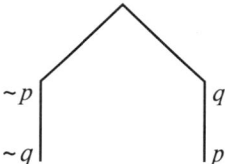

As "predicted", either both p and q are false or both p and q are true.

Testing Arguments by Truth Tree

We are now in a position to see how truth trees can be used to test the validity of arguments.

An argument is valid if it is impossible that its premises be true and its conclusion false. We diagram this *counterclaim*, the conjunction of the premises and the denial of the conclusion. If all the branches of this truth tree are closed, there is no possibility that the counterclaim be true, and the argument is valid. If the tree contains even one open branch, this signifies one way in which the premises might be true and the conclusion false, and the argument is invalid.

It remains to see how to diagram the denial of the conclusion.

Negation does not lend itself to picturing.

Emanuel Gottlieb Leutz painted a picture of Washington Crossing the Delaware. But suppose someone thought that Washington did *not* cross the Delaware and wanted to paint a picture of that. What would he paint?

Instead of trying to work out a way of diagramming denials, we turn to the truth tables for negation that we have already worked out and show them as a list of equivalences:

DN: $\sim\sim p \leftrightarrow p$
NIF: $\sim(p \rightarrow q) \leftrightarrow (p \wedge \sim q)$
NAND: $\sim(p \wedge q) \leftrightarrow (\sim p \vee \sim q)$
NOR: $\sim(p \vee q) \leftrightarrow (\sim p \wedge \sim q)$
NIFF: $\sim(p \leftrightarrow q) \leftrightarrow (p \leftrightarrow \sim q)$
          $\sim(p \leftrightarrow q) \leftrightarrow (\sim p \leftrightarrow q)$
          $\sim(p \leftrightarrow q) \leftrightarrow [(p \wedge \sim q) \vee (\sim p \wedge q)]$

To test an argument for validity, we first restate the premises and the denial of the conclusion in symbolic form. We use the equivalences listed above to *internalize* negation (see the right-hand sides of the equivalences; there the "not" signs are inside) and diagram each statement. Next we diagram the counterclaim; that is, we diagram the conjunction of these statements, for convenience using first the simpler and "narrower" conjunctive statements and those which "mesh" in obvious ways, so as to avoid excessive branching. We then look for an open branch. If there is an open branch, the counterclaim is consistent and the argument is invalid. If every branch is closed, the counterclaim is inconsistent and the argument is valid.

To recapitulate: to build a truth tree for a compound statement or for a conjunction of such statements, we use the following four basic rules:

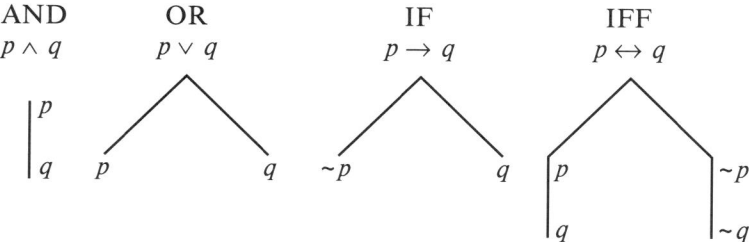

But, first, before these rules can be applied, the statements must be prepared for diagramming by internalizing negation within them, as far as possible, so that the "not" signs apply to single letters only. For this preparation we use the equivalences for negation listed above.

Alternatively, we could use, in addition to the four basic truth-tree rules, Jeffrey's second tier of basic rules, which cover negation of compounds,[5] as follows:

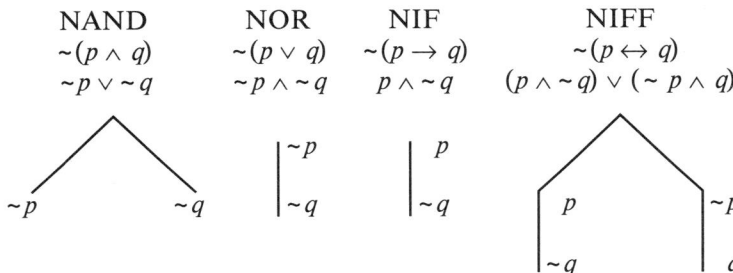

This second set of four rules constitute a convenient short cut, based upon the equivalences for negation and available once those equivalences are clearly understood. Use of these four rules does not make it unnecessary to internalize negation, but it shortens the work. The process of preparing statements for diagramming before they can be entered on a truth tree is of some practical importance, for neither our ordinary-language statements nor their symbolic restatements make immediately clear what their diagrams will look like. The diagramming is not straightforward. Truth trees do not picture directly either conditionals or the denials of compound statements, and so two significant types of replacement are required.

First, since truth-tree branching represents 'or,' not 'if' ('∨', not '→'), when a conditional is to be diagrammed, the antecedent of that conditional must be replaced on the tree by its negation. In effect the equivalence IF is being invoked, and the negation that was implicit in the conditional statement will be made explicit on its truth-tree diagram.

---

[5] See Jeffrey, p. 72.

Secondly, whenever the negation of a compound statement is to be diagrammed it must first be replaced by an equivalent statement in which negation has been fully internalized. In both types of situation the process of preparation for diagramming must be carried out with care.

The equivalences for negation are useful not only for denying the conclusions of arguments, but also for diagramming the premises where these are negative or where they contain negation, explicit or implicit. We saw how this worked in the case of the "exclusive or"; let us take special note of how it works for conditionals with compound antecedents.

In chapter 1, we showed that the argument:

If it rains or snows, school will be closed.
It is raining.
Therefore, school will be closed.

is valid. Let us check this argument by truth tree. First, we formalize. Let $r =$ It is raining. Let $s =$ It is snowing. Let $c =$ School will be closed.

PREM 1: $(r \vee s) \rightarrow c$
PREM 2: r
∴ c

And we formulate the denial of the conclusion:

PREM 1: $(r \vee s) \rightarrow c$
PREM 2: r
~CON: ~c

The first premise contains "implicit" negation, for it is equivalent to ~$(r \vee s) \vee c$. In accordance with the truth-tree rule for IF, the diagram for the first premise will contain two branches, one for the denial of the antecedent, one for the consequent. In order to be clear about what should go on the "antecedent" branch, we refer to the equivalence NOR and see that ~$(r \vee s)$ must be replaced by its equivalent ~$r \wedge$ ~s. Having thus internalized negation, we produce the following truth tree:

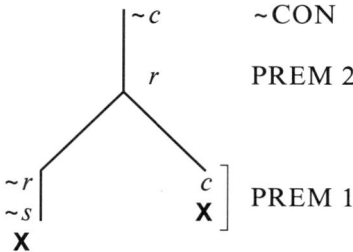

Both branches close, and the argument is valid. Note that the tree is clearly marked, to show where each item comes from. The following diagram would do as well, but is a bit clumsier:

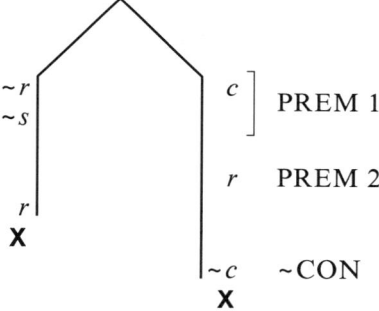

Before going on to discuss *tautologies* – the theorems of truth-functional logic – let us see what the diagram for an *invalid* argument looks like.

First, the *fallacy of affirming the consequent* – called a "fallacy" because it is invalid *and* can be mistaken for modus ponens, which is valid. We show that

$$p \to q$$
$$q$$
$$\therefore p$$

is invalid.

PREM 1: $p \to q$
PREM 2: q
~CON: ~p

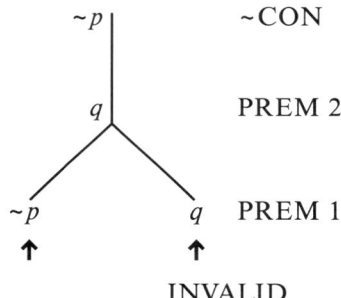

INVALID

where the arrows call attention to the open branches. This form of argument is invalid. The tree shows two open branches; in fact, all its branches are open. But even one open branch is sufficient to show invalidity – we could have stopped half way through the diagramming of the first premise and yet have proved our point:

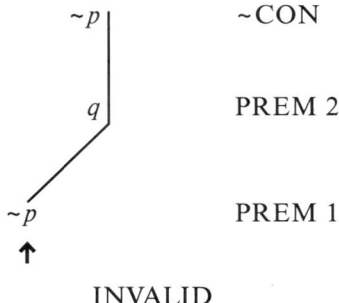

INVALID

Let us look at one more invalid argument, to illustrate a fallacious use of the rule of SIMPlification:

> If I jump off the balcony and wave at my uncle, I will hurt myself. Therefore, if I wave at my uncle, I will hurt myself.

PREM: $(j \wedge u) \rightarrow h$
$\sim(j \wedge u) \vee h$ by IF
$(\sim j \vee \sim u) \vee h$ by NAND
CON: $u \rightarrow h$
~CON: $\sim(u \rightarrow h)$
$u \wedge \sim h$ by NIF

```
       u  ⎤
      ~h  ⎦   ~CON
     /  \
   /      \
  /        \
~j    ~u       h   PREM
↑     X        X
              INVALID
```

Few people would be likely to think that this *example* was valid, but the "other" conclusion: "If I jump off the balcony, I will hurt myself", might well have been more tempting, even though it is equally fallacious, for it is "derived" from the same premise in the same way – by careless misuse of SIMPlification.

A truth tree tests the validity of an argument by testing the consistency of its counterclaim: if the counterclaim – the claim that the premises are true, but the conclusion is false – is inconsistent, the argument is valid. If the counterclaim is consistent, if it can possibly in any way be true, the argument is invalid. (The discovery of one way in which the counterclaim could perhaps be true is the discovery of a counterexample to the argument.)

Testing arguments by truth tree is an *indirect* method. We suppose that the premises of an argument are true and the conclusion false, and we diagram that supposition in detail. If the resulting truth tree closes, with an explicit contradiction on every branch, we have shown the logical impossibility of the supposition, and we have made our point: the argument is valid.

Tautologies

The discussion of chapter 1 relied heavily on a clear distinction between arguments and conditional statements. An argument moves from its premise to its conclusion and claims (among other things) that the conclusion follows reliably from the premise. The corresponding conditional makes only the weaker claim: if the premise, then the conclusion; either in fact the conclusion holds or else the premise does not. The principle of Conditional Proof allows us to use a valid argument to justify the (weaker) claim that its corresponding conditional is true.

The distinction between arguments and statements is reflected in the terminology we have been using. Arguments are said to be valid or invalid. Statements – including conditionals – are said to be true or false. An argument is valid if it meets logical criteria. A statement is true if what it says is so. And to find out whether statements, such as "It is raining" or "If it is raining, the clouds have closed in" are true, we look out the window, not into a logic book.

But the truth tree for an argument forces us to look at its corresponding conditional in a somewhat different way. The counterclaim, by means of which we test the argument, is precisely the denial of the corresponding conditional.

NIF: $(p \wedge \sim q) \leftrightarrow \sim (p \rightarrow q)$

So, if the truth tree shows that the counterclaim is inconsistent and thus that the argument is valid, it shows at the same time that the denial of the corresponding conditional is inconsistent and thus that the conditional itself could not possibly be false. It shows, in other words, not just that the corresponding conditional is true, but that it could not have been otherwise.

It is appropriate, accordingly, to extend the concept of validity to apply to statements as well as arguments. A statement will count as *valid* if there is no possibility that it be false, if it is true no matter what the circumstances, no matter what the truth values of its component parts. (It is still poor usage to speak of an argument as "true" or "false" – it is hard to imagine what that would mean. The premises and conclusion are true or false, but not the argument itself.)

100   Truth-functional Logic

We find it useful to be able to single out those compound statements which are valid. These are the *tautologies* of truth-functional logic, the *theorems* of logic, the "content" – if there is such – of logical culture.

We could use truth tables to discover which compound statements are valid, or tautologous: their tables show 'T' at every row. But let us return, instead, to the method of truth trees.

Testing Statements by Truth Tree

When one is asked whether a given statement is valid, there seems to be a natural impulse – which is to be resisted – to begin by diagramming that statement. Take, for example, the law of the excluded middle, $p \vee {\sim}p$, which we established in chapter 1. Its truth-tree diagram looks like this:

which is extraordinarily uninformative. It tells us only that it is logically possible that $p \vee {\sim}p$, since its branches do not close. Suppose then that we diagram instead $p \vee {\sim}p$ along with its denial ${\sim}(p \vee {\sim}p)$, which, by NOR, becomes ${\sim}p \wedge {\sim}{\sim}p$.

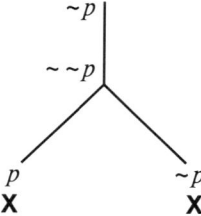

That tree closes; so it tells us something, but not what we wanted to know. It tells us that $p \vee {\sim}p$ and its denial are inconsistent with each other – which is true of *every* statement and its denial. Or, perhaps, if we mark it like this:

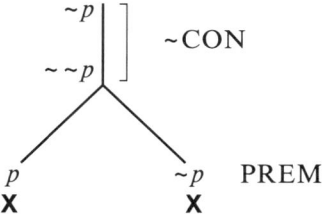

it shows that $p \vee \sim p$ can be validly deduced from itself. Again, a mistaken procedure turns out to be uninformative.

To get it right, we remind ourselves that the method of truth trees is basically indirect, a version of reductio ad absurdum. We diagram – alone – the *denial* of $p \vee \sim p$, $\sim(p \vee \sim p)$, which, by NOR, is $\sim p \wedge \sim \sim p$:

$\sim p$

$\sim \sim p$

X

And that little tree trunk shows that $p \vee \sim p$ is valid! In chapter 1 it took at least eight steps – and some ingenuity – to establish the same result. (It may be well to remember, however, that the truth-tree method has built into it a certain amount of logical information – including seven equivalences for negation.) The method can now be used to test arguments for validity and statements either indirectly for validity or directly for consistency. Truth tables also can be used to test statements for validity and for consistency, though this is often cumbersome.

It should by now be clear to the reader that the two general methods we have been using for establishing the validity of arguments in propositional logic agree; testing and deduction yield the same results. That is, any argument that tests valid by truth tree (or by truth table) can be shown by deduction to be valid, and any argument for which a correct deduction has been given will test valid by truth tree (and by truth table). Likewise, any compound sentence that can be established as a theorem by deduction will turn out to be a tautology (its truth table will show "true" for all truth values of its components, and the truth tree for its negation will close); and vice versa. But this claim – that deduction and testing agree – has by no means been established. It is a claim made not *in* first-order logic, but *about* first-order logic, to be supported by work in "metalogic," at another time, in another place. Like the questions of consistency and completeness which we mentioned in the opening paragraphs of chapter 1, but did not try to handle, this question is best postponed.

Before going on to do problems based on these methods, it may be well to review some of the terminology introduced in this section.

All valid statements (true no matter what, true at every row of their truth tables) are true (true at the row that represents the actual situation in the world, with respect to their component parts). All true statements are consistent (possibly true, true on at least one row). All inconsistent statements (false at every row) are false (false at the row that represents the actual situation in the world); all false statements are invalid (false at at least one row, possibly false).

So all valid statements are consistent (but not vice versa), and all inconsistent statements are invalid (but not vice versa).

Reminders for Chapter 2

Basic Truth Tables:

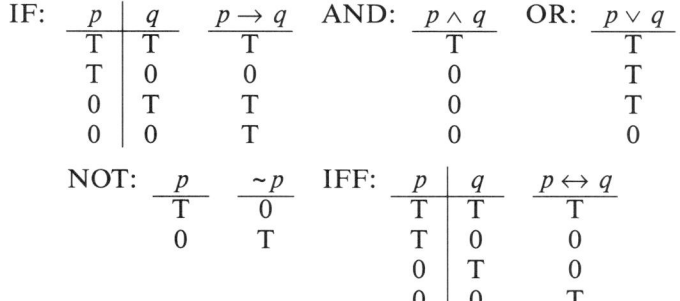

IF:	p	q	$p \to q$	AND:	$p \wedge q$	OR:	$p \vee q$
	T	T	T		T		T
	T	0	0		0		T
	0	T	T		0		T
	0	0	T		0		0

NOT:	p	$\sim p$	IFF:	p	q	$p \leftrightarrow q$
	T	0		T	T	T
	0	T		T	0	0
				0	T	0
				0	0	T

Basic Truth Trees:

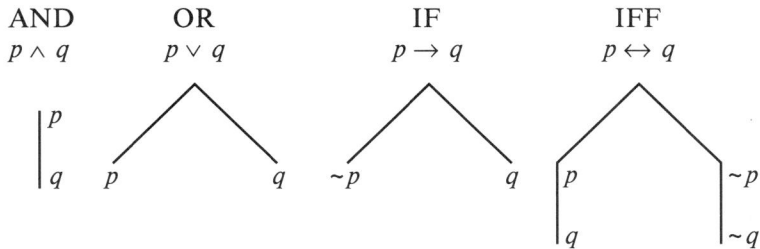

AND OR IF IFF
$p \wedge q$ $p \vee q$ $p \to q$ $p \leftrightarrow q$

Auxiliary Truth-tree Rules

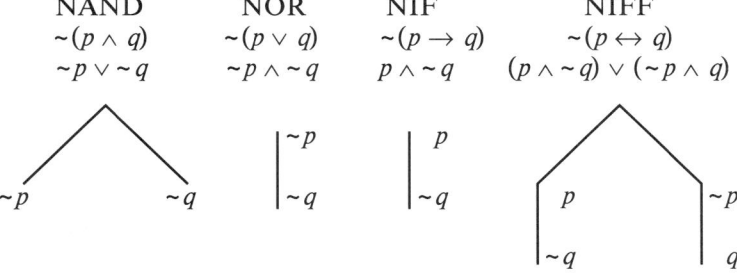

NAND NOR NIF NIFF
$\sim(p \wedge q)$ $\sim(p \vee q)$ $\sim(p \to q)$ $\sim(p \leftrightarrow q)$
$\sim p \vee \sim q$ $\sim p \wedge \sim q$ $p \wedge \sim q$ $(p \wedge \sim q) \vee (\sim p \wedge q)$

Internalizing Negation

The negation of a compound statement can always be replaced by an equivalent statement that is not a negation. We use the following list of equivalences for negation:

DN: $\quad \sim\sim p \leftrightarrow p$

NIF: $\quad \sim(p \to q) \leftrightarrow (p \land \sim q)$

NAND: $\sim(p \land q) \leftrightarrow (\sim p \lor \sim q)$

NOR: $\quad \sim(p \lor q) \leftrightarrow (\sim p \land \sim q)$

NIFF: $\quad \sim(p \leftrightarrow q) \leftrightarrow (p \leftrightarrow \sim q)$
$\leftrightarrow (\sim p \leftrightarrow q)$
$\leftrightarrow [(p \land \sim q) \lor (\sim p \land q)]$

Once negation has been internalized, any compound statement or conjunction of statements can be diagrammed, using only the four basic rules.

On Equivalence and the Biconditional

A *biconditional*, a statement of the form:

so and so if and only if such and such

or $\quad\quad p \leftrightarrow q$

is defined as the conjunction of the two conditionals:

so and so if such and such

$q \to p$

so and so only if such and such

$p \to q$

Accordingly,

$(p \leftrightarrow q) =_{\text{def}} [(p \to q) \land (p \to q)] \quad \text{DEF} \leftrightarrow$

Equivalence is defined as validity of the biconditional. Two statement forms — and, accordingly, two statements — are *equivalent* if it has been *established* (by deduction, by truth table, by truth trees, or however) *as a theorem* that the biconditional holds between them. To establish an equivalence by deduction (or by truth tree), two deductions (or two truth trees) are required, one in each direction, so as to establish two implications, which are then conjoined. To establish an equivalence by truth table, we show only that the tables for the two statements are alike at every row; it is important that the two truth tables be built on the same reference list. Such an equivalence may be cited, if convenient, as a step in proof.

Equivalences for Use in INTerchange

Absorption: $[p \to (p \land q)] \leftrightarrow (p \to q)$

Exportation: $[(p \land q) \to r] \leftrightarrow [p \to (q \to r)]$

Contraposition: $(p \to q) \leftrightarrow (\sim q \to \sim p)$
$(p \to \sim q) \leftrightarrow (q \to \sim p)$

Commutativity: $(p \land q) \leftrightarrow (q \land p)$
$(p \lor q) \leftrightarrow (q \lor p)$

Associativity: $[p \land (q \land r)] \leftrightarrow [(p \land q) \land r]$
$[p \lor (q \lor r)] \leftrightarrow [(p \lor q) \lor r]$

Redundancy: $(p \land p) \leftrightarrow p$
$(p \lor p) \leftrightarrow p$

IF: $(p \to q) \leftrightarrow (\sim p \lor q)$

IFF: $(p \leftrightarrow q) \leftrightarrow (\sim p \leftrightarrow \sim q)$
$(p \leftrightarrow q) \leftrightarrow [(p \land q) \lor (\sim p \land \sim q)]$

Distribution: $[p \land (q \lor r)] \leftrightarrow [(p \land q) \lor (p \land r)]$
$[p \lor (q \land r)] \leftrightarrow [(p \lor q) \land (p \lor r)]$
$[p \to (q \land r)] \leftrightarrow [(p \to q) \land (p \to r)]$
$[p \to (q \lor r)] \leftrightarrow [(p \to q) \lor (p \to r)]$
$[(p \lor q) \to r] \leftrightarrow [(p \to r) \land (q \to r)]$
$[(p \land q) \to r] \leftrightarrow [(p \to r) \lor (q \to r)]$

Problems for Chapter 2

Parts I and II of these problems can be done after the introduction to chapter 2 and the material on truth tables (pp. 4–85) have been read. Part III can be done after the material introducing truth trees (pp. 85–100) has been read, and Part IV at the end, after the section on tautologies.

I. On Negation.

1. Can there be a situation in which a statement and its negation are both true? If so, give an example. If not, explain.

2. Can there be a situation in which a statement and its negation are both false? If so, give an example. If not, explain.

Find the negations of each of the following statements. Begin by formalizing the statement, defining any abbreviations you use (single letters only). Your first statements of each negation will read:

In English,
It is not the case that ...

In symbols,
~[...]

Then use the equivalences for negation to internalize negation, and rewrite the result in ordinary language. You should end up with unambiguous sentences in idiomatic English.

3. Alice will stay home from work only if the rain is very heavy or the road is closed.

4. Mary is interested in music, and Alice is interested in art, but Nora is interested in neither.

5. My uncle George will go to the meeting if and only if my uncle Henry goes.

6. There has been an earthquake in Los Angeles and a flood in St. Louis.

7. If the rain is heavy and Main Street is flooded, the bank will be closed and so will the hardware store.

8. George and Alice both want to go skiing, but George wants to go and Alice does not.

II. Use truth *tables* to demonstrate three of the equivalences shown in the Reminders for this chapter:

1. Absorption: $[p \to (p \land q)] \leftrightarrow (p \to q)$
2. Contraposition: $(p \to q) \leftrightarrow (\sim q \to \sim p)$
3. Distribution: $[p \land (q \lor r)] \leftrightarrow [(p \land q) \lor (p \land r)]$

Use truth tables also to demonstrate the following theorems:

4. $(p \land \sim p) \to q$
5. $(p \land q) \to (p \lor q)$
6. $(p \to r) \to [(p \land q) \to r]$

III. Test the following arguments for validity by truth *tree*. If you have any question about the correctness of your results, check them by truth *table*. Do at least one truth-table check.

1. If Peter gives a party Quentin or Rose will come.
   If Quentin and Rose come, there will be group singing.
   Therefore, if Peter gives a party there will be group singing.

2. If Peter gives a party, Quentin and Rose will come.
   If Quentin or Rose comes, there will be group singing.
   Therefore, if Peter gives a party, there will be group singing.

3. If Peter goes to Albany, he will visit his grandmother.
   If Peter goes to New York, he will march in the parade.
   Therefore, Peter will either visit his grandmother or march in the parade.

4. If Peter goes to Albany he will visit his grandmother.
   If he does not go to Albany, he will march in the parade.
   Therefore, Peter will either visit his grandmother or march in the parade.

5. Henry was born in Philadelphia or in San Francisco.
   Henry was born in San Francisco.
   Therefore, he was not born in Philadelphia.

6. Henry was not born in both San Francisco and Philadelphia.
   Henry was born in San Francisco.
   Therefore, he was not born in Philadelphia.

7. If Alice goes both to the grocery store and to the butcher shop she will get everything she needs.
   Alice will go to the butcher shop if she goes to the grocery store.
   Therefore, if Alice goes to the butcher shop she will get everything she needs.

8. If Henry is to please his grandmother he must get A in French.
   If he is to get A in French he will not have time for basketball.
   If he has no time for basketball he will be depressed.
   If he is depressed he will not please his grandmother.
   Therefore, he will not get A in French.

9. Henry will please his grandmother only if he gets A in French.
   If he gets A in French he will not have time for basketball.
   He will be depressed if he has no time for basketball.
   If he is depressed he will not please his grandmother.
   Therefore, he will not please his grandmother.

10. If George does not lend Henry $50 if Henry asks him to, Henry will be disappointed.
    Henry will not be disappointed.
    Therefore, Henry will ask George to lend him $50.

11. Henry will be disappointed if George does not lend him $50 if Henry asks him to.
    Henry will not be disappointed.
    Therefore, George will lend Henry $50.

12. The government statistics are reliable and the cost of living has gone down or the newspaper reports are reliable and unemployment has gone up.
    Neither the government statistics nor the newspaper reports are reliable.
    Therefore, economics is confusing.

13. If the government statistics are reliable the cost of living has gone down, and if the newspaper reports are reliable unemployment has gone up.
    Neither the government statistics nor the newspaper reports are reliable.
    Therefore, the cost of living has not gone down and unemployment has not gone up.

IV. Use truth trees to decide whether the following statements are tautologous (truth-functionally valid). Check your answers by truth table.

1. It is not the case that if it is raining it is not raining.

2. It is not the case that it is raining and not raining.

3. If school will be closed if it snows, then school will not be closed if it does not snow.

4. Either school will be closed if it snows or school will not be closed if it does not snow.

5. Either school will be closed if it snows or school will not be closed if it snows.

6. I won't wear my hat if I take my umbrella if and only if I won't take my umbrella if I wear my hat.

Chapter 3
Evaluating Arguments

We have seen that a proposed argument can be approached in at least two different ways: by deduction (chapter 1) or by a checking procedure (chapter 2). One can take seriously the premises of the argument, suppose that they are true, and try to deduce the proposed conclusion from them in a step-by-step rule-governed way. Or one can examine the argument statically, and use some systematic checking procedure (truth tables, truth trees, or some other method) to find out whether it is logically possible that the premises be true but the conclusion false. The checking procedure we have recommended in chapter 2 is indirect: we assume the counterclaim, diagram that assumption on a truth tree, and see whether there is an open branch – whether it is indeed possible that the premises be true and the conclusion false. Using either truth tables or truth trees and working systematically, we can find a counterexample to any proposed argument or theorem, if indeed there is one.

These two different approaches have certain advantages and disadvantages. Deduction is illuminating in that it makes clear why an argument works if it does work, what principles of inference it invokes. Deduction relies on human ingenuity for devising proofs and so trains the mind. In addition, there are complex arguments that are not open to test[1]; such arguments can be handled systematically only by deduction. But deduction has the obvious disadvantage that it does not establish invalidity – one can waste a great deal of time and energy trying to prove things

---

[1] No such arguments occur in either propositional or (monadic) predicate logic, that is, in the material covered in Parts I and II of this book.

that are not so. Checking procedures, where they are available, avoid that sort of waste. They also avoid another kind of waste: the use of human ingenuity where thoughtless routine would do. Mechanical procedures can yield yes-or-no answers as to whether arguments are valid. And the method of truth trees, being indirect, may be informative as well, in that, where a given argument is *in*valid, its truth tree will produce a counterexample. Indirect testing procedures can make clear to us why an argument does *not* work if it does not work, what "loopholes" it leaves open.

Accordingly, in view of the various advantages and disadvantages of the two approaches, we shall use them in tandem. We shall at the same time attend to one of the long-term purposes of studying logic in the first place: to learn to recognize, promptly and with some degree of reliability, a valid argument or an invalid argument when we see one. So we shall assess each argument initially, before testing and before any attempt at deduction, to see whether it makes sense to us. Afterward, we shall use truth trees and (perhaps) deduction to check our intuitions. Studying specific arguments should be a learning exercise not only in deductive and testing techniques, but, more importantly, in reasoning.

Exhibiting Structure

A first step in assessing an argument is to exhibit its structure. This is the point of the use of symbols. So far, we have introduced five symbols for five truth-functional connectives: →, ∧, ∨, ~, and ↔; we have used '∴' for "therefore", to introduce the conclusion of an argument; and we have used single-letter abbreviations for sentences. That is enough for now. (We have also numbered our premises, numbered the steps of a deduction, and introduced a star system to keep track of premises in play – but these are matters of bookkeeping, which have little to do with the initial assessment of an argument.)

We begin by defining our abbreviations – explicitly if there is any possibility of ambiguity. Each single letter abbreviates a whole sentence – even if the original English-language presentation of the argument "telescopes" – omits parts of sentences for purposes of grouping – or uses names in one place and pronouns in another. For example, in problem III.1 of chapter 2 we have the premise: "If Peter gives a party, Quentin or Rose will come." Appropriate abbreviations would be:

Let p = Peter gives a party.
Let q = Quentin comes to the party.
Let r = Rose comes to the party.

yielding, as restatement of the first premise, '$p \to (q \vee r)$'. Here 'p' does *not* stand for "Peter" or for "party," nor 'q' for "Quentin" nor 'r' for "Rose"; each letter stands for a *statement*, as listed above. With 's' for "There will be group singing," we rewrite the argument as follows:

$$p \to (q \vee r)$$
$$\underline{(q \wedge r) \to s}$$
$$\therefore p \to s$$

As the reader has probably discovered for herself, III.1 turns out to be invalid. That is, we can find a counterexample, a situation in which the premises would be true but the conclusion false. The premises leave open the possibility that Peter might give a party to which Quentin comes but not Rose (or Rose but not Quentin) and there is no group singing; in that case the conclusion would be false.

If the argument in question had been the following:

If Peter gives a party, Quentin and Rose will come. If Quentin and Rose come to the party, there will be group singing. Therefore, if Peter gives a party, there will be group singing.

which is simpler in that the consequent of its first premise and the antecedent of its second premise are alike, the abbreviations could have been different, so as to reflect that simplicity. They might be:

Let p = Peter gives a party.
Let w = Quentin and Rose come to the party.
Let s = There will be group singing.

– three letters, rather than the four needed in chapter 2.
In other words, the abbreviations selected must be sufficient to expose the structure of the argument, in this case:

$$p \to w$$
$$w \to s$$
$$\therefore p \to s$$

but should not, ordinarily at least, introduce needless complexity. Our simpler argument is, of course, obviously valid, an instance of the chain rule.

One caveat: even when it would seem the obvious thing to do, never include a negative in an abbreviation. This is for practical reasons, not a matter of principle. If 'q' is used to stand for "Quentin does not play tennis," it is extraordinarily easy to forget the "not," use 'q' somewhere in place of "Quentin does play tennis," and get badly mixed up.

Having set up whatever abbreviations are needed, we are now in a position to discover the logical structure of the argument, on the basis of which the argument will either succeed or fail. We now replace the "little words," the "logical words," of the informally stated argument by the appropriate truth-functional connectives, paraphrasing as needed and deliberately ignoring any rhetorical flourishes that may obscure the structure. Whereas the connectives used in ordinary language are various – many of them alike or almost alike in meaning – and occasionally ambiguous, the logical connectives, which are expressed in symbols, are few and unambiguous. This is the revolutionary contribution of modern symbolic logic. Use of abbreviations and variables is old, at least as old as Aristotle; use of symbols for clearly defined logical concepts is relatively new, initiated in the seventeenth century by Leibniz and developed in the nineteenth century by Boole, De Morgan, C.S. Peirce, and others. As a result, we now have a way to articulate the structure of our arguments, for purposes of logical evaluation and understanding. Learning to do this properly requires practice and discernment.

We have already seen that we must be careful in symbolizing a conditional; the point is to be clear about which is the antecedent and which the consequent. "Only if" is not to be confused with "if" or with "if and only if." Since it means "if not q then not p," 'p only if q' goes into '$\sim q \to \sim p$' or into its contrapositive '$p \to q$' indifferently – the choice is a matter of taste or convenience. And logical order is not to be confused with linguistic order; for example, 'q, if p' is the same as 'if p, q'; i.e., '$p \to q$'.

One "rhetorical flourish" that is to be ignored is the emphasis on contrast conveyed by such words as 'although', 'while', 'whereas', 'nevertheless', or 'but'. The reader may perhaps have noticed that the locutions 'the premises are true but the conclusion false' and 'the premises are true and the conclusion false' have occurred frequently and indifferently in these pages. 'But' and 'and' as sentential connectives are interchangeable for our purposes; both are symbolized by '$\wedge$'.

'Unless' is synonymous with 'if not'. Many logic books recommend reading 'unless' as 'or'; this is convenient, but it is a bit of logical sophistication which should not be taken immediately for granted. Let us take the trouble to show that ($\sim q \to p$ and $p \vee q$ have the same truth table and so are genuinely interchangeable. (An example: you will be punished unless you are quiet; you will be punished if you are not quiet; if you are not quiet you will be punished; you will be quiet or you will be punished; you will be punished or you will be quiet.)

p	q	$\sim q$	p	$\sim q \to p$	$p \vee q$
T	T	0	T	T	T
T	0	T	T	T	T
0	T	0	0	T	T
0	0	T	0	0	0

The reader is now free to read 'unless' as 'or' or 'if not' as he pleases.

Non-truth-functional Connectives

There are a number of sentential connectives and compounds that cannot be handled truth-functionally. One is the counterfactual conditional (if p had been so, q would have been so), about which there is an extensive philosophical literature.[2] A *counterfactual* (or contrafactual) conditional is a "conditional" in which the antecedent is known to be false and the consequent is stated in the subjunctive mood; for example,

> If Eleanor Roosevelt had lived in eighteenth-century China, her feet would have been bound.
>
> If Leigh Cauman had been a boy, she would not have attended the Horace Mann School for Girls.
>
> If silicon had been a gas, I would have been a major general.[3]
>
> If Bizet and Verdi had been compatriots, Bizet would have been Italian.
>
> If Bizet and Verdi had been compatriots, Verdi would have been French.

(The last two are Quine's examples.[4])

The point of such a statement is to claim that what is conveyed by the consequent is causally or otherwise contingent on the truth of the antecedent. In other words, a counterfactual conditional *asserts* the very causal or other connections we have been at pains to eliminate from consideration. If such a statement is misinterpreted as an ordinary conditional it must be taken to be true, regardless of the consequent, without further inquiry, because the antecedent is false. This is misleading. Counterfactual conditionals should not be interpreted as conditionals, any more than accused criminals should be treated as criminals.

For related reasons, 'because' and its synonyms, even when they function as sentential connectives, cannot be thought of as truth-functional. "I passed the course because I gave an apple to the teacher" is obviously false if I did not give an apple to the

---

[2] Nelson Goodman, "The Problem of Counterfactual Conditionals," *Journal of Philosophy*, vol. 44 (1977), pp. 113-128, variously reprinted and commented on. Also, *Methods*, part I, ch 3, and Jeffrey, *op.cit.*, pp. 50-52.

[3] This is a quote from James McNeill Whistler, in conversation, referring to the fact that he flunked out of West Point in 1852.

[4] *Methods*, 1st ed., p. 15; 4th ed., p. 23.

teacher and obviously false if I did not pass, but, for the case in which both components are true, the would-be builder of a truth table is at a loss, for he has insufficient basis for deciding the truth value of the compound. In the context of first-order propositional logic, sentences of the form "p because q" must be symbolized – if at all – by a single letter, say, 'b'.

'Since', unlike 'because', usually does not occur as a sentential connective. Rather, it introduces a premise of an argument, just as 'therefore' or 'so' introduces a conclusion.

'Either' tends to be misleading in a different way. It is often thought that it connotes the exclusive "or" and so affects the sense of the sentences in which it occurs. This is seldom the case. The sense of the exclusive "or" – symbolized '$(p \vee q) \wedge \sim(p \wedge q)$' – is usually conveyed by context or gesture or by raising one's voice, rather than by the word 'either'. ("You may have ice cream *or* cake, and that's that," for example. "You may have ice cream or cake, but not both" would be clearer.) The function of 'either ... or' (like that of 'both ... and') is usually to bracket components of a compound sentence, for purposes of grouping.

The observant reader will already have realized that parentheses: ( ) – and brackets: [ ], and braces: { }, as well – are used in logic as they are in algebra, for grouping or packaging, to make clear the structure of a compound. In ordinary language parentheses are more likely to be used to set off asides – phrases or sentences that interrupt the flow of the discussion. See the paragraph before this one for examples of both uses. See also the paragraph on parentheses in the Reminders to this chapter.

Quine[5] disambiguates the sentence:

John will play or John will sing and Mary will sing.

by telescoping:

John will play or sing and Mary will sing.
John will play or John and Mary will sing.

---

[5] *Methods*, part I, ch. 4, and Answers to Exercises.

Disambiguation can also be accomplished (much less elegantly in this case) by insertion of 'either' or 'both', perhaps associated with changes of word order:

Mary will sing and either John will play or John will sing.
John will play or both John will sing and Mary will sing.

'Either ... or' serves to bracket what comes between the two words, marking that material as the left-hand component of a disjunction. Similarly, 'both ... and' brackets the left-hand component of a conjunction. Clearly, neither device can indicate where the disjunction or the conjunction ends.

'Neither ... nor' brackets similarly. It also makes clear one odd "fact" of language: that the logical equivalence NOR – in strong contrast with NAND and NIF – is built into our ordinary usage. Attending only to the words, one would expect that "Neither Mary nor George will go" ("Neither will Mary go nor will George go) would be read intuitively as the denial of a disjunction: "Not either Mary or George will go," "Not either Mary will go or George will go," "It is not the case that either Mary will go or George will go," in symbols, '$\sim(m \vee g)$'. In fact, "Neither Mary nor George will go" is more likely to be read immediately as "Mary will not go and neither will George" or "Mary will not go and George will not go," where the main connective is "and," in symbols, '$\sim m \wedge \sim g$'. Since NOR tells us that these readings are equivalent, either one will do.

Finding the Main Connective

In complicated sentences given in ordinary language it is sometimes not immediately clear which is the main connective and which are the subsidiary ones. This is a matter of some importance, since the main connective of a sentence defines its over-all structure – whether it should be classified as a conditional, a negation, a conjunction, a disjunction, or a biconditional – and so determines which of the principles of inference are relevant to it.

Modus Ponens, for instance, tells us that a conditional – a

statement whose main connective is '→' – can be used by taking it as a premise, together with its antecedent, and deducing its consequent as a conclusion. Conditional Proof tells us that, conversely, a conditional can be established by taking its antecedent as premise and deducing its consequent.

A conjunction – a statement whose main connective is '∧' – can be used by SIMPlification or established by ADJunction.

A disjunction – a statement whose main connective is '∨' – can be used by DILemma or ELIMination or established by THINning.

A negation – a statement whose main "connective" is '~' – can be established by REDuctio ad absurdum and can be used in many ways, among them DN, Modus Tollens, and ELIMination.

It is useful, accordingly, to be quite clear about which are the main connectives in the sentences that enter into the arguments we are trying to understand.

Let us reconsider an argument from chapter 1 (see pages 33 ff.).

If Henry is accepted in the program, he will be studying economics and he will be studying hospital management.

Therefore, Henry will be studying hospital management and if he is accepted in the program he will be studying economics.

Let us suppose that we take a quick look at this argument and guess that it is valid, on the grounds that the premise and conclusion seem to say, roughly, the same thing. Let 'a' stand for "Henry is accepted in the program"; let 'e' stand for "Henry will be studying economics"; and let 'h' stand for "Henry will be studying hospital management." We take the premise to be a conjunction, its main connective, "and." So we formalize:

$$\frac{(a \to e) \land h}{\therefore h \land (a \to e)}$$

This argument is obviously (even trivially) valid, by Commutativity of '∧', confirming our first guess.

But there is another way of looking at the premise. We could take it to be a conditional with conjunctive consequent, its main connective "if," not "and." In that case the formalized argument looks like this:

$$\frac{a \to (e \wedge h)}{\therefore\ h \wedge (a \to e)}$$

For this argument to be valid each of the conjunctive components of its conclusion must follow separately from the premise. Perhaps $a \to e$ follows from $a \to (e \wedge h)$, but h certainly does not, since a conditional does not guarantee the truth of either its antecedent or its consequent. So the argument is invalid. We confirm this judgment by truth tree:

PREM: $\sim a \vee (e \wedge h)$ by IF
$\sim$CON: $\sim[h \wedge (a \to e)]$
         $\sim h \vee \sim(a \to e)$ by NAND
         $\sim h \vee (a \wedge \sim e)$ by NIF

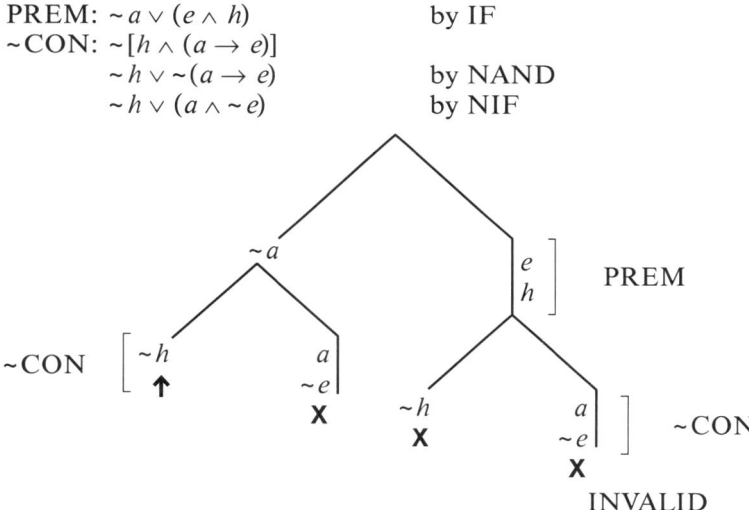

The moral of this tale is that we must be very careful in determining which are the main connectives in the statements we are dealing with, for such determinations are consequential. In this example the premise as stated above was ambiguous. That ambiguity is easily avoidable in English, by punctuation perhaps, or by telescoping:

> If Henry is accepted in the program he will be studying economics and hospital management.

which is how the premise was presented in chapter 1, where the aim was to illustrate the fallacy of illicit repetition in the proposed deduction.

In responsible argument, in context, such ambiguities are to be avoided. Unfortunately, ordinary language almost inevitably involves ambiguities; correct formal notation does not. The careful use of brackets and parentheses allows us to avoid them (see the memos on parentheses in the Reminders for this chapter and for chapter 1).

Evaluation of Arguments

We turn now to the actual assessment of arguments. We will deal with this briefly; for, like the recognition of structure, assessment is largely a matter of practice and discernment – the province of the student, not of the teacher. Some suggestions, however, are worth making.

First, it is often useful to reorganize an argument (at least in one's head) into a familiar form or into a form that one happens to like. Some students like disjunctive elimination; some like chain rule; some like dilemma; others like reductio. Sometimes this reorganization can be accomplished just by changing the order of the premises, sometimes by using one or another of the familiar equivalences: contraposition, absorption, IF, NIF, etc. For example, given an argument like this:

If Joe knew about the meeting, he planned to come.
Joe did not plan to come or he missed his train.
If Joe missed his train, he will come on the late bus.
Therefore, either Joe did not know about the meeting or he will come on the late bus.

$$1 \quad k \to p$$
$$2 \quad {\sim}p \lor t$$
$$3 \quad t \to b$$
$$\therefore \quad {\sim}k \lor b$$

one can recognize in it an argument of this form:

$\sim p \vee t$	PREM 2
$\sim p \rightarrow \sim k$ (If Joe did not plan to come, he did not know about the meeting.)	PREM 1, by Contraposition
$t \rightarrow b$	PREM 3
$\therefore \sim k \vee b$	

which looks like a dilemma and is recognizably valid. Or again, one could reorganize the same argument like this:

$k \rightarrow p$	PREM 1
$p \rightarrow t$	PREM 2, by IF
$t \rightarrow b$	PREM 3
$\therefore k \rightarrow b$	CON, by IF

which is a valid chain. Either reorganization would make the validity obvious and the assessment easy.

Secondly, it is sometimes possible to discern "gaps" in argument and determine what would be needed to fill those gaps. If what is needed can be provided legitimately, the argument is valid, otherwise not. For an example, let us go back to III.1 of the problems of chapter 2:

$$p \rightarrow (q \vee r)$$
$$\underline{(q \wedge r) \rightarrow s}$$
$$\therefore p \rightarrow s$$

What is missing for the construction of a successful chain is a "middle" premise:

Added:
$$p \rightarrow (q \vee r)$$
$$(q \vee r) \rightarrow (q \wedge r)$$
$$\underline{(q \wedge r) \rightarrow s}$$
$$\therefore p \rightarrow s$$

Inserting that middle premise results in a chain argument which is valid. But we have been given no reason to believe that if either Quentin or Rose goes to that party, both Quentin and Rose will go. (Here the words 'either' and 'both' have been used for emphasis, not for grouping.) So the needed insertion is illegitimate, and the argument being evaluated is invalid.

Argument III.2, however, looked like this:
$$p \to (q \wedge r)$$
$$\underline{(q \vee r) \to s}$$
$$\therefore p \to s$$

The extra premise needed to fill the gap in this argument was $(q \wedge r) \to (q \vee r)$, which is tautologous. If Quentin and Rose both go to the party, surely at least one of them will go.

Added (theorem)
$$p \to (q \wedge r)$$
$$(q \wedge r) \to (q \vee r)$$
$$\underline{(q \vee r) \to s}$$
$$\therefore p \to s$$

So argument III.2 is valid.

Roughly the same contrast occurred in III.3 and III.4.

III.3 $a \to g$
$$\underline{n \to p}$$
$$\therefore g \vee p$$

III.4 $a \to g$
$$\underline{\sim a \to p}$$
$$\therefore g \vee p$$

Each argument needed a first premise, a disjunction, to yield a familiar dilemma form. III.3 needed $a \vee n$. III.4 needed $a \vee \sim a$. There was no reason to assume $a \vee n$; $a \vee \sim a$ is tautologous. So III.3 is invalid, and (but?) III.4 is valid. Presumably the reader has already found that the truth-tree method confirms these results.

Note that, characteristically, an invalid argument is inconclusive: the premises do not provide enough information to guarantee the truth of the conclusion. An invalid argument is not like a mistake in arithmetic, which can be corrected by replacing a wrong answer with the right one. More often than not, what is wrong when we use an invalid argument is not that we have drawn the wrong conclusion, but that we have drawn a conclusion from premises that do not warrant any (nontrivial[6]) conclusion at all.

---

[6] A "trivial conclusion" in this context is a conclusion that could be drawn from one premise alone, without the others.

We noted earlier that one advantage of the method of truth trees is that, when it shows that an argument is invalid, it produces at the same time a counterexample. Let us go back once again to problem III.1. Its truth tree might look like this:

PREM 1: $p \to (q \lor r)$
PREM 2: $(q \land r) \to s$
  CON: $p \to s$
 ~CON: $p \land {\sim}s$ by NIF

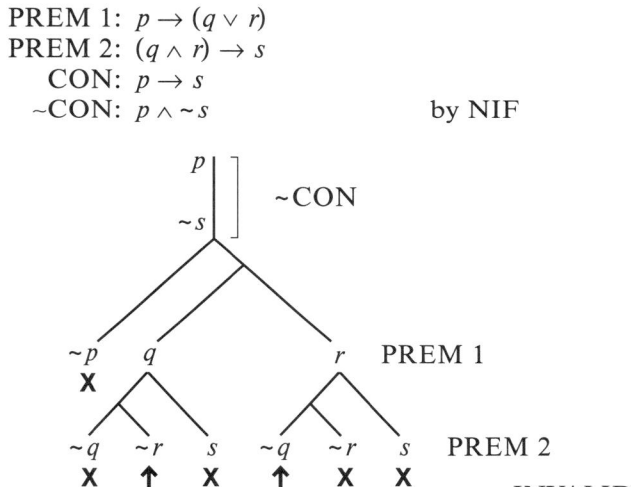

There are two open branches, from which we can read off the two counterexamples: $p \land {\sim}s \land q \land {\sim}r$ and $p \land {\sim}s \land r \land {\sim}q$. The conclusion of the argument can fail, because the premises leave open the possibility that Quentin can come to the party without Rose or Rose without Quentin.

The method of truth trees has one further advantage, which may be of interest, again deriving from the focus on inconsistency. In testing one argument with two premises, we are really testing three arguments at once: the original argument and two other arguments, each of which has as premises one of the original premises and the denial of the original conclusion and which has as conclusion the denial of the other original premise. In discovering that three statements are inconsistent (or consistent), we discover that *any* two of them imply (or fail to imply) the denial of the other one.

In other words, the tree we have just exhibited shows the invalidity of three arguments: the one we started with and also:

If Peter gives a party Quentin or Rose will come. (P1)
Peter will give a party but there will be no
  group singing. (~CON)
Therefore, Quentin and Rose will come
  to the party, but there will be no group singing. (~P2)

Peter will give a party at which there is no
  group singing. (~CON)
If Quentin and Rose come to the party,
  there will be group singing. (P2)
Therefore, Peter will give a party, but neither
  Quentin nor Rose will come. (~P1)

The same two counterexamples (one would of course suffice) defeat all three arguments. Given the premises of each argument, it is logically possible that Peter will give a party, there will be no group singing, and one of Quentin and Rose will come but not the other; in both of those cases the conclusion will be false. The key to the invalidity of all three arguments is that Quentin or Rose's coming to the party does not guarantee that both will come – nor (equivalently) does Quentin or Rose's staying away guarantee that neither will come.

The economy involved in testing three arguments for validity at one time – put more generally, the economy of testing for inconsistency rather than testing for validity – has been of theoretical interest at various points in the history of philosophy. It is the basis, for example, for the theory of the antilogism, devised by Christine Ladd-Franklin at the turn of the century, to replace the cumbersome rules then used for the testing of categorical syllogisms.[7] Professor Ladd-Franklin correlated each syllogism with its counterclaim, or *antilogism*, the conjunction of its two premises with the denial of its conclusion, and then designed an extremely simple test for deciding whether such an antilogism was indeed an inconsistent triad. If so, the original syllogism was valid. An extension of Ladd-Franklin's antilogism can be seen in

---

[7] Christine Ladd-Franklin, "On the Algebra of Logic," in Charles Sanders Peirce, ed.: *Studies in Logic* by members of the John Hopkins University (Boston: Little Brown, 1883).

Quine's indirect method for testing predicate-logical arguments in the first and second editions of *Methods*. Testing for inconsistency rather than for validity often turns out to effect interesting economies. And recognizing the inconsistency or consistency of a counterclaim is sometimes easier than recognizing the validity or invalidity of an argument.

Another strategy in the assessment of arguments, which is sometimes useful, is to strip away what appears to be irrelevant detail, or clutter, so as to get at the central argument, and assess that. This strategy, of course, must be used with care, for what seems at first to be irrelevant detail may turn out to be important.

An example may be in order.

My aunt Ellen is looking for her glasses. If she cannot find them (f) she cannot read the telephone book (r), and if she cannot read the telephone book she cannot phone her sister in Los Angeles (s). Well, she says to herself, I used my glasses this morning to read the newspaper in the kitchen at breakfast (u), and either I put them away properly in my glasses case (c) or I am wearing them (w) or I left them in the kitchen (k). Here is the empty glasses case (e); I did not put them away. So they must be in the kitchen. Good, there they are, on the kitchen table; I have found them. So I can phone my sister.

There is a bit of narrative in the above, and two arguments, both defective, but in different ways.

First:
$$\frac{u \wedge (c \vee w \vee k)}{\therefore \ k}$$
$$e \wedge \sim c$$

Disregarding u and e as redundant and irrelevant, we obtain the central argument:
$$\frac{c \vee w \vee k}{\therefore \ k}$$
$$\sim c$$

This argument works by ELIMination, and it felt valid in the context of the narrative. But inspection of the structure shown above makes quite clear that it is *not* valid, for Aunt Ellen has not mentioned that she does not have her glasses on. The truth tree for the full argument confirms this:

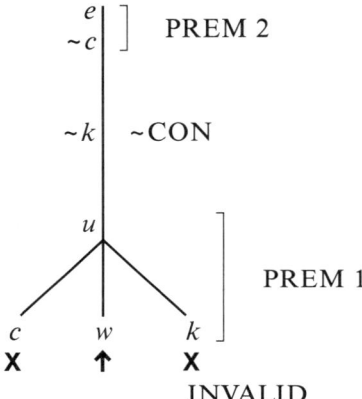

INVALID

Now we can correct this argument by adding a missing premise: ~w – noting that Aunt Ellen knew she did not have her glasses on, even if she didn't say so, even to herself. This is something the Greeks might have dignified by the title "enthymeme" – a "valid" argument depending essentially upon a missing premise that is "understood." Similar arguments may not be so easily repaired: missing premises may be unavailable, and important. When one uses a disjunction in argument, whether by DILemma or by ELIMination, one must be careful to deal with every disjunctive component of the disjunction.

The second argument is also patently invalid.

$$\begin{array}{r}\sim f \to \sim r \\ \sim r \to \sim s \\ \underline{k \wedge f} \\ \therefore \ s\end{array}$$

Noting the chain argument of the first two premises and the irrelevance of k, we condense it to:

$$\begin{array}{r}\sim f \to \sim s \\ \underline{f} \\ \therefore \ s\end{array}$$

This central argument is obviously fallacious, an instance of the *fallacy of denying the antecedent*, the corresponding fallacy of Modus Tollens. Again, a truth tree (or truth table) for the full argument will confirm this assessment.

Strategy in Deduction

In closing this chapter, I would like to review some of the material in chapter 1, which the reader will be using in doing the problems for chapter 3. These problems call for deductions in just those cases in which the statements or arguments have tested valid. The nine basic principles of inference suggest four useful strategies:

(1) To prove a conditional, take its antecedent as premise and try to prove its consequent.
(2) To prove the denial of a statement, take the statement itself as premise and try to derive a direct contradiction.
(3) To use a disjunction, take as premises each of its disjuncts separately and try to derive a joint conclusion.
(4) When in doubt, try reductio. Take as premise the direct denial of what you want to prove, and plan to use DN with reductio at the end.

But it should be recognized always that deduction is a matter of human ingenuity. There are always many ways in which a deduction can be done, and choosing among them is, in part at least, a matter of taste. In particular, the four auxiliary rules: chain rule, modus tollens, nonconjunction, and disjunctive elimination, should not be forgotten. Elimination, for example, is often more convenient than dilemma.

It should be remembered, also, that, once a deduction has been done, it need not be done again; so special attention should be paid to the two principles of inference that are specifically designed to facilitate the use of logical information already obtained. These principles are THEOREM and INTerchange. A short list of theorems and a longer list of equivalences appear among the Reminders for chapter 3. The reader may wish to add to these lists other items that she herself has established.

The rule THEOREM allows the arguer to insert as a step in proof a substitution instance of any theorem already established. Occasionally it is convenient to cite as a theorem a tautology that one is certain *is* a theorem and establish it separately on the side. When cited at the outset of a proof, a theorem carries no stars, but, when it is cited in the course of a proof, it carries the stars appropriate to its position. No other steps of the proof in which it is being used should be cited as justification for a theorem.

The principle of INTerchange authorizes the interchange of equivalents at any point in a deduction. Although the specific equivalence that justifies such an interchange is, of course, a theorem, it does not ordinarily appear *as a step* in the deduction. But it is helpful to the reader of the deduction if we cite that equivalence, along with the principle of INTerchange, if possible by name (see step 6 below). The result of INTerchange follows from an earlier step that is exactly like it except for the interchange; that earlier step must be in play and must be cited.

Here is a sample deduction that makes use of both devices:

If today is Wednesday or Friday, Mary has a music lesson.
If Mary has a music lesson or an appointment with the dentist, she cannot play basketball.
Therefore, if Mary is playing basketball, today is not Friday.

$*\begin{cases} 1 \\ 2 \end{cases}$ $(w \vee f) \to m$ PREM
      $(m \vee d) \to \sim b$

To Prove: $b \to \sim f$

In designing this deduction we will use the principle of INTerchange even before we set to work. We note that the contrapositive of the desired conclusion – that is, "If today is Friday, Mary will not play basketball" – seems to follow straightforwardly from the premises. So our strategy will be to prove that statement first, that is, to prove: $f \to \sim b$, which is equivalent to $b \to \sim f$ and so INTerchangeable with it (see steps 5 and 6 below).

$*\begin{cases} 1 & (w \vee f) \to m \\ 2 & (m \vee d) \to \sim b \end{cases}$ PREM

To Prove: $f \to \sim b$

* 3 $f \to (w \vee f)$ THEOREM
* 4 $m \to (m \vee d)$ THEOREM
* 5 $f \to \sim b$ 3, 1, 4, 2 CHAIN
* 6 $b \to \sim f$ 5 INT
                                 (Contraposition, DN)
                                 QED

This obviously valid argument could have been shown to be valid by a deduction that used only the nine basic rules – but such a deduction would have been much longer. The reader may prefer a different pattern of argument, using, perhaps, reductio or modus tollens instead of CHAIN. For example, following an over-all strategy of conditional proof,

$*\begin{cases} 1 & (w \vee f) \to m \\ 2 & (m \vee d) \to \sim b \end{cases}$ PREM

To Prove: $b \to \sim f$

** 3 b PREM (for CP)
** 4 $\sim \sim b$ 3 DN
** 5 $\sim (m \vee d)$ 2, 4 MT
** 6 $\sim m \wedge \sim d$ 5 INT (NOR)
** 7 $\sim m$ 6 SIMP
** 8 $\sim (w \vee f)$ 1, 7 MT
** 9 $\sim w \wedge \sim f$ 8 INT (NOR)
** 10 $\sim f$ 9 SIMP
* 11 $b \to \sim f$ 3–10 CP
                            QED

Reminders for Chapter 3

Principles of Inference

Modus Ponens: From a conditional statement taken together with its antecedent the consequent follows as a conclusion.

Conditional Proof: From a correct derivation of a conclusion from a premise (or set of premises) the corresponding conditional: if the premise(s) then the conclusion, follows as a summary conclusion.

SIMPlification: From a conjunction any of its conjunctive components follows.

ADJunction: From any statements in play at any point in a deduction their conjunction follows.

REDuctio ad absurdum: From a correct derivation of a direct contradiction from a premise, the denial of that premise follows as a summary conclusion.

Double Negation (DN): A statement and its double negation are interchangeable, whether as steps in a deduction or as components of such steps.

DILemma: Given a disjunction, if each of its disjuncts has been shown by a separate deduction (lemma) to lead to the same conclusion, that conclusion follows from the disjunction.

THINning: From any one (or more) of the disjunctive components of a disjunction the disjunction itself follows.

REPEAT: In the course of a deduction it is always legitimate to repeat as a step in proof any premise that is still in play or any step in proof whose premise is still in play.

CHAIN: From a pair (or sequence) of conditional statements in which the consequent of one conditional is the antecedent of the next, the conditional statement that has the first antecedent as its antecedent and the last consequent as its consequent, follows as a conclusion.

Modus Tollens: From a conditional statement taken together with the denial of its consequent the denial of the antecedent follows.

NonCONJunction: From the denial of a conjunction, taken together with a component of the conjunction, the denial of the other conjunct(s) follows.

ELIMination: From a disjunction, taken together with the denial of a disjunctive component, the rest of the disjunction follows.

THEOREM: Any substitution instance of a theorem, once established, may be introduced as a step in proof at any point in a deduction.

INTerchange: Any pair of statements that have been shown to be equivalent, or deducible one from the other, may be interchanged, whether as steps in deduction or as components of such steps.

Basic Truth Tables

IF:	p	q	$p \to q$	AND:	$p \wedge q$	OR:	$p \vee q$
	T	T	T		T		T
	T	0	0		0		T
	0	T	T		0		T
	0	0	T		0		0

NOT:	p	$\sim p$	IFF:	p	q	$p \leftrightarrow q$
	T	0		T	T	T
	0	T		T	0	0
				0	T	0
				0	0	T

Basic truth trees

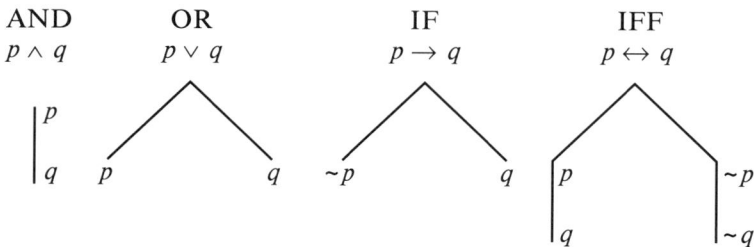

The items entered on a truth tree are sentence letters or the negations of sentence letters, only. See above.

To diagram the *negation* of a compound statement, replace it by an equivalent statement with negation *internalized*; use one or more of the equivalences for negation. Diagramming requires that negation, even implicit negation (see the diagrams for "if" and "iff" above), be fully internalized, so that the "~" signs apply only to single letters.

A truth-tree branch containing a sentence letter and *its* negation is *closed*; we mark this closing with an **X**. A branch that contains no such contradiction is *open*; such a branch may be marked with an arrow. To check whether a branch is open, we trace the branch all the way from its ending to the top of the tree, disregarding other branches. In entering a statement on a truth tree, we must be certain to enter a diagram for that statement on each of the open branches of the tree.

A statement or conjunction of statements is *consistent* if its truth-tree diagram contains at least one open branch, *inconsistent* if all its branches are closed.

To test an *argument* for validity, diagram its *counterclaim*: the conjunction of its premises and the denial of its conclusion. If this truth tree closes, the counterclaim has been shown to be inconsistent, and the argument is valid. If the truth tree contains even one open branch, the counterclaim is consistent, and the argument is invalid. To understand *why* an argument is invalid if it is, we reexamine an open branch of the truth tree and read off from it a set of circumstances under which the premises of the argument would be true but its conclusion false.

To test a *statement* for validity, we diagram the *denial of that statement*. If this truth tree closes, the statement is valid; if it contains an open branch, the denial is consistent, and the statement is invalid.

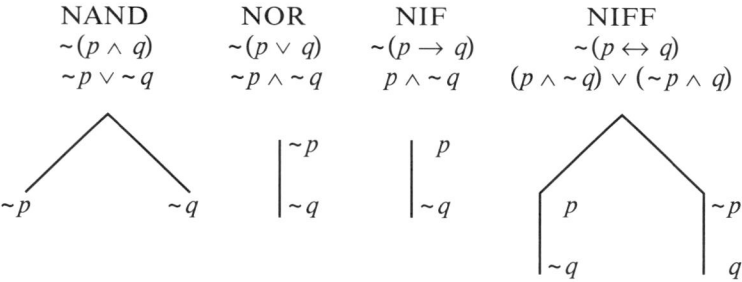

Auxiliary Truth-tree Rules

Equivalences for Negation

DN: $\sim\sim p \leftrightarrow p$

NIF: $\sim(p \to q) \leftrightarrow (p \wedge \sim q)$

NAND: $\sim(p \wedge q) \leftrightarrow (\sim p \vee \sim q)$

NOR: $\sim(p \vee q) \leftrightarrow (\sim p \wedge \sim q)$

NIFF: $\sim(p \leftrightarrow q) \leftrightarrow (p \leftrightarrow \sim q)$
$\leftrightarrow (\sim p \leftrightarrow q)$
$\leftrightarrow [(p \wedge \sim q) \vee (\sim p \wedge q)]$

Other Equivalences

Absorption: $[p \to (p \wedge q)] \leftrightarrow (p \to q)$

Exportation: $[(p \wedge q) \to r] \leftrightarrow [p \to (q \to r)]$

Contraposition: $(p \to q) \leftrightarrow (\sim q \to \sim p)$
$(p \to \sim q) \leftrightarrow (q \to \sim p)$

Commutativity: $(p \wedge q) \leftrightarrow (q \wedge p)$
$(p \vee q) \leftrightarrow (q \vee p)$

Associativity: $[p \wedge (q \wedge r)] \leftrightarrow [(p \wedge q) \wedge r]$
$[p \vee (q \vee r)] \leftrightarrow [(p \vee q) \vee r]$

Redundancy: $(p \wedge p) \leftrightarrow p$
$(p \vee p) \leftrightarrow p$

IF: $(p \to q) \leftrightarrow (\sim p \wedge q)$

IFF: $(p \leftrightarrow q) \leftrightarrow (\sim p \leftrightarrow \sim q)$
$(p \leftrightarrow q) \leftrightarrow [(p \wedge q) \vee (\sim p \wedge \sim q)]$

Distribution: $[p \wedge (q \vee r)] \leftrightarrow [(p \wedge q) \vee (p \wedge r)]$
$[p \vee (q \wedge r)] \leftrightarrow [(p \vee q) \wedge (p \vee r)]$
$[p \to (q \wedge r)] \leftrightarrow [(p \to q) \wedge (p \to r)]$
$[p \to (q \vee r)] \leftrightarrow [(p \to q) \vee (p \to r)]$
$[(p \vee q) \to r] \leftrightarrow [(p \to r) \wedge (q \to r)]$
$[(p \wedge q) \to r] \leftrightarrow [(p \to r) \vee (q \to r)]$

Theorems

$p \to p$ law of identity

$p \to \sim \sim p$ double negation

$\sim \sim p \to p$ double negation

$\sim (p \wedge \sim p)$ noncontradiction

$p \to (q \to p)$

$\sim p \to (p \to q)$

$(p \wedge \sim p) \to q$

$(p \wedge q) \to q$
$p \to (p \vee q)$
$(p \to r) \to [(p \wedge q) \to r]$
$p \vee \sim p$ excluded middle

The reader may wish to add other theorems to this list.

On Parentheses

Parentheses are used for grouping; they tell us to think of what is between them as a package, as a whole. They are needed to dispel ambiguity. They are not needed when there is no ambiguity.

Because '$(p \wedge q) \wedge r$' is equivalent to '$p \wedge (q \wedge r)$' and '$(p \vee q) \vee r$' is likewise equivalent to '$p \vee (q \vee r)$', parens in those cases are unnecessary; they can be dropped in favor of '$p \wedge q \wedge r$' and '$p \vee q \vee r$', respectively. Parens are not needed in strings of letters connected uniformly by '$\wedge$' (and) or '$\vee$' (or).

Parentheses are needed, however, whenever the arrow for "if" is repeated. '$p \to q \to r$' is quite meaningless, since $(p \to q) \to r$ and $p \to (q \to r)$ are significantly different. And '$p \to q \to r$' does *not* mean "if p then q and if q then r" on the misleading analogy of '$x < y < z$' in algebra or '$2 < 4 < 9$' in arithmetic. '$p \to q \to r$' is just bad grammar.

Parens are also needed whenever propositional connectives are mixed. "We will go to the movies if it rains, and we will have dinner together" is different in meaning from "If it rains we will go to the movies and have dinner together." In ordinary language this difference is sometimes conveyed by a difference in word order or by some other linguistic device; in symbols it is conveyed by a difference in parentheses, e.g.,

$(r \to m) \wedge d$ versus $r \to (m \wedge d)$

The ambiguous '$r \to m \wedge d$', like '$p \to q \to r$', is always disallowed as ill-formed, ungrammatical. Likewise, '$p \wedge q \vee r$'.

Without parens the "not" sign '$\sim$' applies to the single letter that follows it; with parens it applies to the whole grouping (compound sentence) that immediately follows it. Since '$\sim \sim p$' means '$\sim(\sim p)$' unambiguously, parens are not needed in that case.

Otherwise, in formalizing compound sentences, parentheses to indicate grouping are always needed. Unnecessary parens are occasionally a nuisance, but not a mistake. Missing parens that lead to ambiguity do constitute a mistake.

I have used square brackets throughout to cover expressions that already contain parentheses. This practice makes for ease in reading, but it is not a matter of principle.

Problems for Chapter 3

Part I of these problems should be done after the first ten pages of this chapter have been studied. After this whole short chapter has been read the other problems are in order – or a selection from them – as well as any problems postponed from chapter 1. This is an opportunity to review all of Part I of the book before going on to Part II.

I. Exhibit the structure of the following arguments. Define explicitly all the abbreviations you use (single letters only).

1. Alice can get fresh fish for dinner only if she goes to the fish monger on Main Street.
Alice cannot go both to the grocery store and to the fish monger.
Therefore, Alice cannot get fresh fish unless she goes to the fish monger but not to the grocery.

2. Mary will have to take a leave of absence next semester and earn some money unless she can get an increase in her scholarship grant.
Getting an increase in scholarship money will require significant improvement in her grades.
Therefore, if Mary continues to care more about athletics than about studying and her grades do not improve, she will have to take a leave of absence.

3. George will graduate in May only if he satisfies the science requirement.
He will satisfy the science requirement if and only if he gets at least a B in logic.
Therefore, if George does not get at least a B in logic or if he fails logic altogether, George will not graduate in May.

4. If Joe's parents like the idea of Joe's going to medical school and can raise the money to send him, Joe will go to medical school.
If Joe's parents don't like the idea of Joe going to medical school but he can get a scholarship, Joe will go to medical school.
Joe is assured of a scholarship if his parents cannot raise the money.
Therefore, Joe will go to medical school, whether his parents like it or not.

II. Assess the validity of the four arguments of Problem I. That is, with respect to each of them, decide whether the argument seems valid to you. Then check your judgment by truth tree. If the argument is valid, show that it is valid by formal deduction; and, if not, explain what is wrong with it.

III. Consider the following arguments. With respect to each,

(1) exhibit its structure; define your abbreviations unless they are quite obvious;
(2) decide whether or not the argument seems valid to you;
(3) check the validity of the argument by truth *tree*;
(4) if the argument is valid, show that it is by formal deduction; and if not, explain why not.

1. If Joe's parents like the idea of his going to medical school they will raise the money to send him and he will be able to go.
Joe will get a scholarship if his parents don't want him to go, and will go to medical school anyway.
Therefore, whether his parents like it or not, Joe is going to medical school.

2. If Henry gets up early and has a good breakfast, he will get to work on time.
   Henry will have a good breakfast only if he gets up early.
   Therefore, Henry will get to work on time only if he gets up early.

3. If Henry gets up early and has a good breakfast, he will get to work on time.
   Henry will have a good breakfast if he gets up early.
   Therefore, if Henry gets up early, he will get to work on time.

4. If Henry gets up early and has a good breakfast, he will get to work on time.
   Henry will have a good breakfast only if he gets up early.
   Therefore, if Henry gets up early, he will get to work on time.

5. Henry will get to work on time if he gets up early and has a good breakfast.
   Henry will have a good breakfast.
   Therefore, if he gets up early, he will get to work on time.

6. If there is a snow storm or a holiday, parking restrictions will be lifted.
   If parking restrictions are lifted, you can park your car on our street tomorrow morning.
   You cannot park on the street tomorrow.
   Therefore tomorrow is not a holiday.

7. Albert will fail algebra if Mary does not help him with his algebra problems.
   If Albert fails either algebra or geometry he will be in trouble.
   But if Mary does help Albert with his algebra, she will not have time for her own studying and she will be in trouble.
   Therefore, either Albert or Mary will be in trouble.

8. Joe will not get a license to drive unless he is over eighteen years old and passes the driving test.
   Joe is twenty years old and a very good driver.
   If Joe takes the driving test he will pass it.
   Therefore, if Joe takes the driving test, he will get his driver's license.

9. We are going hiking, and if the sun shines we will enjoy ourselves.
   Therefore, if the sun shines we will go hiking and enjoy ourselves.

10. If the sun shines we will go hiking and enjoy ourselves.
    Therefore, if we go hiking and the sun shines we will enjoy ourselves.

11. If we go hiking and the sun shines we will enjoy ourselves.
    Therefore, if the sun shines we will go hiking and enjoy ourselves.

12. If we go hiking and the sun shines we will enjoy ourselves.
    Therefore, we are going hiking, and if the sun shines we will enjoy ourselves.

13. If the sun shines we will go hiking and we will enjoy ourselves.
    Therefore, we will enjoy ourselves and if the sun shines we will go hiking.

14. Alice will get to the meeting on time only if she drives dangerously fast.
    If Alice drives dangerously fast she will be stopped by the policeman on Main Street and she will not get to the meeting on time.
    Therefore, Alice will not drive dangerously fast.

15. Alice will get to the meeting on time only if she drives dangerously fast.
    If Alice drives dangerously fast she will be stopped by the policeman on Main Street and taken to the police station.

If Alice is taken to the police station, she will not get to the meeting on time.
Therefore Alice will not get to the meeting on time.

16. If I have not taken both algebra and the calculus, I can take neither geology nor physics.
I have taken algebra and I am taking geology.
Therefore, I can take physics.

17. If I have not taken both algebra and the calculus, I can take neither geology nor physics.
I have taken algebra and I am taking geology.
Therefore, I have taken calculus.

18. If he has offered a warranty, the manufacturer can be held responsible for defective operation of your radio if and only if you have mailed in your registration.
The manufacturer cannot be held responsible.
Therefore, even though the manufacturer did offer a warranty, you did not mail in your registration.

19. If he has offered a warranty, the manufacturer can be held responsible for defective operation of your radio if and only if you have mailed in your registration.
The manufacturer cannot be held responsible.
Therefore, the manufacturer did not offer a warranty or you did not mail in your registration.

20. Grocery store rents will go up and so will grocery store prices if rent controls are lifted.
Grocery store rents will go up unless the mayor intervenes.
Therefore, if rent controls are lifted grocery prices will go up unless the mayor intervenes.

21. If no new price controls are imposed then if grocery store rents go up, prices for groceries will go up as well.
Grocery store rents will go up unless the mayor intervenes.
Therefore, if no new price controls are imposed grocery prices will go up unless the mayor intervenes.

22. If rent controls are lifted, rents will go up and so will prices.
If new price controls are imposed, rents will go up but grocery store prices will not.
If the mayor intervenes, new price controls will go into effect and rent controls will stay in place.
Therefore, if the mayor intervenes, neither rents nor grocery prices will go up.

23. If rent controls are lifted, rents will go up and so will prices.
If new price controls are imposed, rents will go up but grocery store prices will not.
If the mayor intervenes, new price controls will be imposed and rent controls will not be lifted.
Therefore, grocery prices will go up only if the mayor does not intervene.

24. If rent controls are lifted, rents will go up and so will prices.
If price controls are imposed, prices will not go up, but rents will.
If the mayor intervenes, rent controls will be lifted and price controls will be imposed.
Therefore, the mayor will not intervene.

IV. Exhibit the structure of the following statements and then, with respect to each of them, decide whether you think it is tautological (truth-functionally valid). Check your judgment by truth tree. If you find that the statement is tautological, establish it by deduction.

1. If it is raining and snowing then it is raining or snowing.

2. If it is not both raining and snowing then it is neither raining nor snowing.

3. If it is neither raining nor snowing then it is not both raining and snowing.

4. If it is neither raining nor snowing then it is both not raining and not snowing.

5. If the statement is not a theorem, its truth tree will have an open branch and its truth table will have a row marked false, or, if the statement is a theorem, its truth tree will be closed and a deduction can be designed.
6. If the statement is not a theorem its truth tree will have an open branch and its truth table will have a row marked false, and, if the statement is a theorem, its truth tree will be closed and a deduction can be designed.
7. If I enjoy tennis and if I am good at tennis if and only if I enjoy it, then I must be good at tennis.

V. In part III of the problems for chapter 1 six equivalences useful for INTerchange were assigned for deduction by means of the nine basic rules of inference. The following three distribution equivalences were omitted from that list:

1. $[p \vee (q \wedge r)] \leftrightarrow [(p \vee q) \wedge (p \vee r)]$
6. $[(p \rightarrow q) \vee (p \rightarrow r)] \leftrightarrow [p \rightarrow (q \vee r)]$
8. $[(p \rightarrow r) \vee (q \rightarrow r)] \leftrightarrow [(p \wedge q) \rightarrow r]$

Establish each of these three, first by truth *table* and then by deduction. One direction (as shown here, from left to right) can be done by DILemma, using the basic rules. Working in the opposite direction (here, from right to left), feel free to use also INT, THEOREM, and the four auxiliary rules, with any theorems or equivalences that you yourself have already established.

Part II
Predicate Logic

Chapter 4
Principles of Inference

The notion of generality has been central to all the work we have done so far. A logical principle is acceptable only if it works every time it is invoked and is reliable in one case just as it is in another. We look at a truth-functional argument: if it meets general criteria for the validity of truth-functional argument, this argument is valid; otherwise not. We look at a compound sentence: if all sentences that share its structure are tautologous, this statement is tautologous; if all statements that share its structure are inconsistent, this statement is inconsistent; and so forth. But what do we mean by saying that two arguments or two statements have "the same structure"?

As we have seen, we can make explicit the truth-functional structure of an argument (or of a compound statement) by finding and abbreviating its component sentences and then first paraphrasing its ordinary-language connectives and then replacing them with unambiguous symbolic connectives from a very short list. Two arguments (or two statements) have the same structure if their symbolic connectives are precisely alike and if their propositional abbreviations can be correlated one by one.

To take a simple and familiar example, the following two arguments have the same structure:

If the moon is made of green cheese, I'll eat my hat.	$(m \rightarrow h)$
I will not eat my hat.	$(\sim h)$
Therefore, the moon is not made of green cheese.	$(\sim m)$

George will celebrate tonight if he gets A in Latin.	$(a \rightarrow c)$
George is not going to celebrate tonight.	$(\sim c)$
So I guess he didn't get A in Latin.	$(\sim a)$

This particular structure has a name, 'modus tollens'. It guarantees that these two arguments are valid.

A key factor in recognizing the structure of an argument is noticing which abbreviations within it match. In both of the arguments above, the proposition denied by the second premise is precisely the consequent of the first premise, and the proposition denied by the conclusion is precisely the antecedent of the first premise. So both arguments are *instances* of modus tollens. They are alike in structure. Since modus tollens has been shown to be an acceptable auxiliary rule of inference, all its instances – these included – are valid.

Until now we have neither examined nor formalized the relation of a general claim to its instances. This is the province of quantification theory.

The need to formalize concepts of generality becomes immediately clear if we pay attention to a number of very familiar structures of argument which we are all quite ready to evaluate but which cannot be handled by the propositional logic we have been studying. For example,

Everything is interesting
Therefore, this book is interesting.
(Valid)

This book is boring.
Therefore, everything is boring.
(Invalid)

All doctors are well educated.
Everyone who is well educated has read *Ulysses*.
Therefore, all doctors have read *Ulysses*.
(Valid)

Some doctors are well educated.
Some who are well educated have read *Ulysses*.
Therefore, some doctors have read *Ulysses*.
(Invalid)

All doctors are well educated.
Everyone who has read *Ulysses* is well educated.
Therefore, some doctors have read *Ulysses*.
(Invalid)

If someone has a gun, Mary is frightened.
Therefore, if Joe has a gun, Mary is frightened.
(Valid)

If everyone has a gun, Mary is frightened.
Therefore, if Joe has a gun, Mary is frightened.
(Invalid)

In the history of logic a segment of quantification theory – the logic of the Aristotelian syllogism – was developed earlier than propositional logic, which was developed by the Stoics. But modern logic builds the logic of quantification from a basis in the logic of propositions, reversing the historical order. This seems to me to make good sense, and I am presenting the material in that way. We shall make minimal use of the traditional Aristotelian methods and vocabulary.

We begin by introducing some new notation, needed in order to articulate the internal structure of a sentence that expresses a general claim. We then take note of two basic (and familiar) principles of quantification theory, one moving from the general and universal to the specific, the other from the specific to the general but existential. We next introduce a second pair of principles of inference, complementary to the first, and a pair of equivalences (provable and so redundant, but important) for quantificational negation. Testing procedures we will leave to the next chapter.

A *universal* claim is a claim that such and such (is foolish, is fallible, is fascinated by French motion pictures, makes funny noises every time it sees a mouse, has forty-four windows, is a ferryboat, is food for thought, or whatever) is true of *everything*. To express such a claim we use the prefix '$\forall x$' (or '$\forall y$') an upside-down A (for "all") with a variable (x, y, z, or any of these with primes or subscripts) – read "for all x" or "whatever x you pick" – taken together with an *open sentence*[1], say '*Fx*', or "*x* is such and such," with both prefix and open sentence clearly marked by parentheses (or brackets).

An *existential* claim is a claim that such and such is true of *something or other*. To express such a claim we use the prefix '∃*x*' (or '∃*y*', etc.): a backwards E with a variable, attached to an open sentence containing an occurrence (or occurrences) of that variable, again with both prefix and open sentence clearly marked by parentheses.

The notion of a variable will be familiar to the reader from elementary algebra or science. A *variable* is a symbol (usually '*x*' or '*y*' or a Greek letter) used in place of a noun or verb or some longer expression, in order to express generality, to make clear that, in the context in question, it does not matter which specific item we are talking about. I have been using 'such and such' and 'so and so' in just that way. The primary use (though by no means the only use, or even the most important use) of variables is in the articulation of rules and definitions. For example,

Let "*x* is a mammal" be short for "*x* is an animal that suckles its young."

An *open sentence* is something that looks and sounds like a sentence, but fails to be one because it is incomplete: it contains at least one occurrence of a variable in place of a noun phrase or verb phrase.[2] An open sentence without a quantifying prefix

---

[1] I have used the term 'open sentence' for what Whitehead and Russell called a "propositional function." See Alfred North Whitehead and Bertrand Russell, *Principia Mathematica* (Cambridge: University Press, 1970; 2nd ed. 1935; hereafter, PM), Introduction; see especially p. 15.

[2] An open sentence may contain occurrences of two variables or more. We deal in Part II with the logic of open sentences (predicates) in one variable only, monadic predicate logic, and postpone until Part III the logic of open sentences in two variables or more, polyadic predicate logic, or the logic of *relations*. We shall occasionally need '*y*' as well as '*x*', even in Part II, but this will be because of nesting of quantifiers, not because of the complexity of the predicates.

In presenting predicate logic separately, before the logic of relations, I am following the policy of W.V. Quine, both in his teaching and in his published pedagogical works; see, e.g., *Methods*.

makes no claim; it is not a sentence; it is incapable of truth or falsity. The variable is not a noun or a verb; it is a blank, a mechanism for cross reference, more convenient than '--' or '...' would be, but no different in meaning. It is more like a pronoun than like any other part of speech, the major difference between them being that whereas a pronoun in ordinary language can hold its reference indefinitely, depending upon the skill of the speaker and the memory of the listener, a variable has no such flexibility. Unless the context is quite clear in specifying otherwise, there are no significant connections between occasions of use.

A quantifying prefix, or *quantifier*, does specify otherwise. Within the *scope* of the quantifier, that is, within the open sentence to which the quantifier applies, the variable of quantification holds its reference. So '$(\forall x)(Fx \vee Gx)$' can abbreviate "whatever item x you pick, it will turn out to be frightening, or else it – that very item – will be agreeable," i.e., "Everything is either frightening or agreeable." And '$(\forall y)(Fy \to Gy)$' can abbreviate "For any y, if y is a ferryboat, then that y is green," or "All ferryboats are green." Within the scope of a quantifier the variable of quantification holds its reference – but not beyond it. If the same variable occurs again, without a quantifier, even in the same argument or the same sentence, it does not achieve cross reference – it "dangles." We shall say that a quantifying prefix *binds* those occurrences of its variable of quantification which are within its scope, and that other occurrences of that variable, outside that scope and not so bound, are *free*. So '$(\forall x)(Fx) \vee Gx$' is an open sentence, since the 'x' in 'Gx' is free. '$(\forall x)(Fx) \vee Gx$' is interchangeable with '$(\forall y)(Fy) \vee Gx$'.

'Fa', on the other hand, does not abbreviate an open sentence, and neither does '$(\forall x)(Fx) \vee Fa$'. For we *stipulate* that 'a', 'b', 'c', etc., the lower-case letters at the beginning of the alphabet – 'x', 'y', 'z', etc. having been set aside for variables – will stand for real names or referring phrases or *pseudonames*, names introduced ad hoc for the purposes at hand. (The notion of a pseudoname we will discuss later in this chapter.) So 'Fa' can stand for, say, "Alice is foolish" or "Alfred plays football," and '$(\forall x)(Fx) \to Fa$' can stand for, say, "If everyone likes fireworks, my aunt Amy likes fireworks," which is surely true.

The following argument, accordingly, is valid on truth-functional grounds (we are here using quantificational *notation*, not quantificational principles of inference):

Everything is either beautiful or ugly, or Alice is much mistaken.
Alice is not mistaken.
Therefore everything is beautiful or ugly.

$$(\forall x)(Bx \vee Ux) \vee Ma$$
$$\underline{\sim Ma}$$
$$\therefore (\forall x)(Bx \vee Ux)$$

The above is valid, by ELIMination. The following (simpler) notation would have been sufficient for showing the validity of the argument:

$$E \vee A$$
$$\underline{\sim A}$$
$$\therefore E$$

Two Unrestricted Rules

Quantificational reasoning is needed, on the other hand, for this argument:

Everything is beautiful or ugly.
Therefore, Alice is beautiful or ugly.

$$\underline{(\forall x)(Bx \vee Ux)}$$
$$\therefore Ba \vee Ua$$

The principle invoked here is *Universal Instantiation* (UI, specification, or $\forall$ elimination):

Universal Instantiation: From a universally quantified statement any of its instances, without restriction, follows.

Notationally, an instance of a quantified statement is the result of dropping the quantifier and substituting a referring expression for the variable of quantification in the associated open sentence.

Accordingly, $(\forall x)(Fx)$
               $\therefore Fa$ UI

UI is a *principle* of inference; so '*Fx*' can be replaced by any open sentence, however simple or complex, regardless of content. We have already seen that UI applies to truth-functional compounds of open sentences (which are also open sentences): '*Fx* ∨ *Gx*', '*Fx* → *Gx*', etc. We have also seen that the truth-functional principles of inference (from chapter 1) apply to the fully formed statements of quantification theory (so far, universal claims and singular statements like *Fa*), but not to its open sentences, which are incapable of truth or falsity.

The principle of INTerchange, on the other hand, does apply to these open sentences, and authorizes the use, in quantificational contexts, of all the truth-functional equivalences that we have accepted or established. So, for example, by the equivalence IF [(*p* → *q*) ↔ (~*p* ∨ *q*), which has (*Fa* → *Ga*) ↔ (~*Fa* ∨ *Ga*) as an instance], '*Fx* → *Gx*' is interchangeable with '~*Fx* ∨ *Gx*'. Accordingly, $(\forall x)(Fx \to Gx)$ is equivalent to $(\forall x)(\sim Fx \vee Gx)$ – say, "All birds have wings" is equivalent to "Everything either is not a bird or else has wings."

Also, since variables are merely vehicles for cross reference, sentences alike except for having different variables of quantification are interchangeable. '$(\forall x)(Fx)$' is interchangeable with '$(\forall y)(Fy)$' and, as we shall see below, '$(\exists z)(Gz)$' with '$(\exists w)(Gw)$'. One must be careful, of course, to be consistent in such matters: '$(\forall x)(Fx \to Gx)$' is interchangeable with '$(\forall y)(Fy \to Gy)$', but not with '$(\forall z)(Fz \to Gy)$'.[3]

One must also be careful to use UI *as stated*, not "freely" as one might use a rule of interchange. UI authorizes the dropping of an outside universal quantifier *where that quantifier covers a whole sentence*. That is, it authorizes the inferences:

---

[3] Both of these claims about interchange can be demonstrated metalogically, but we shall just use them. I make the perhaps unjustified assumption that they are intuitively acceptable. They are restated on page 167 and in the Reminders for this chapter.

$(\forall x)(Hx \to Wx)$ — All who are honest are welcome.
∴ $Ha \to Wa$ — ∴ If Albert is honest, he is welcome.

and

$(\forall x)(\sim Hx)$ — Everyone (everything) is dishonest.
Nobody (nothing) is honest.
∴ $\sim Ha$ — ∴ Albert is not honest.

But UI does not authorize the (fallacious) inferences:

$(\forall x)(Hx) \to Ss$ — If everyone is honest, Samuel will be surprised.
∴ $Ha \to Ss$ — ∴ If Albert is honest, Samuel will be surprised.

or

$\sim(\forall x)(Hx)$ — Not everyone is honest.
∴ $\sim Ha$ — ∴ Albert is not honest.

Nor does UI authorize the following inferences, which are valid, but which must be shown to be so in other ways:

$(\forall x)(Wx) \vee \sim R$ — Everything is wet or it is not raining.
∴ $Wa \vee \sim R$ — ∴ Alice is wet or it is not raining.

$W \wedge (\forall x)(Mx)$ — The war is over and everything is a mess.
∴ $W \wedge Mb$ — ∴ The war is over and this building is a mess.

One further point should be made before we move on to existential generalization. UI is unrestricted. This means that if $(\forall x)(Fx)$ is true – say, if everything is food for thought – then Alice Jones is food for thought, the pigeon on the ledge is food for thought, and so is the cloud on the horizon and your embarrassment and my sore tooth. And if $(\forall x)(Fx \to Gx)$ is true – say, if all ferryboats are green – then if Alice Adams is a ferryboat, Alice Adams is green, and if that pigeon is a ferryboat, the pigeon is green. Furthermore, and more interestingly, given $(\forall x)(Fx)$, we can derive Fa even when we do not know (or care) what 'a' stands for, when it could be *anything*. '$(\forall x)(Fx)$' tell us that "for all x, whatever you pick as x, x will be F". And surely you can pick something; any old thing will do.

But can you? Perhaps there is nothing at all. The rule UI makes sense only if there is something in the universe to which its universally quantified statements can apply. We have come again upon a metaphysical assumption [cf. chapter 1, p. 35, the law of noncontradiction: $\sim(p \wedge \sim p)$] – this time: "There is something." Again, I cannot argue for the truth of this assumption; it seems to me obviously true, on the face of it. But I think it is important to notice that the assumption is being made and that it functions in our work.[4] We should also notice how weak the assumption is: it does not tell us that there are ferryboats or fishes or clouds or laws or even human beings. (We may of course decide to postulate or claim the existence of such items, separately and explicitly, if we have reason to.) But all that is needed *for the principle of universal instantiation* is that there be something, that our reasoning occur within a nonempty universe.

The second kind of general claim we shall be dealing with is existential, the claim that something, rather that everything, is so and so. Like universal claims, existential claims are nonspecific, general, and this generality is again expressed notationally through the use of variables. We use the existential prefix: a backwards E, '∃', with a variable, attached to an open sentence in that variable, to convey the claim that there is at least one item x such that x is F, or I can pick an x such that x is F, or something is F.

Universal statements can be thought of as long conjunctions: $(\forall x)(Fx)$ tells us that $Fa \wedge Fb \wedge Fc \wedge \ldots$, where the list of items that are said to be F is unmanageably long, comprising everything in the universe. Existentials can be thought of similarly as long disjunctions: '$(\exists x)(Fx)$' tells us that $Fa \vee Fb \vee Fc \vee \ldots$ So UI is an analogue of SIMPlification (though no more than an analogue): it allows us to deduce from a universal any of its instances, or "con-

---

[4] See for example, Karel Lambert and Bas C. van Fraassen, *Derivation and Counterexample* (Encino, Calif.: Dickenson Publishing Company, 1972), ch. 4, sec. 5, and ch. 9. Lambert and Van Fraassen use a system ("free logic"; see p. 129) that does *not* assume that "there is something" (see p. 95).

junctive" components. And the principle of *existential generalization* (EG, quantificational thinning or weakening, or ∃ introduction) is an analogue of THINning: it allows us to deduce an existential from any of its instances, its "disjunctive" components.

Existential Generalization: An existentially quantified statement follows from any of its instances.

Accordingly,

$$\frac{Fa}{\therefore (\exists x)(Fx)} \quad \text{EG}$$

The Addison Gallery has forty-four windows; therefore, something has forty-four windows. Alice likes French movies; therefore something (or, more politely, someone) likes French movies. And so forth. Like UI, EG is unrestricted. Unlike UI, it carries no metaphysical baggage.

One must be careful to understand correctly such a statement as "Some birds have wings"; it tells us that there is something with two properties: being a bird and having wings. It does not tell us merely that there is something that has wings *if* it is a bird. After all, it is true of me (Leigh Cauman, retired logic teacher, woman, not bird) that if I am a bird I have wings (see Part I), or $Bc \to Wc$. You know this is so because I have told you that I am not a bird, and so, since I am not a bird, you know that if I am a bird, I have wings. From this it follows, by EG, that $(\exists x)(Bx \to Wx)$. And you would not be likely to produce me – or this fact about me – as evidence for the view that some birds have wings. Whereas "All birds have wings" is formalized as '$(\forall x)(Bx \to Wx)$', "Some birds have wings" must be formalized as '$(\exists x)(Bx \wedge Wx)$'.

One must also be careful not to be misled by quirks of language. Such sentences as

Someone who goes out in this weather is quite mad.
He who laughs last laughs best.
A boy scout is courteous.

are all *universals*.

One must be careful finally to distinguish clearly between specific, or *singular*, statements and the existentials derivable from them, which are nonspecific and so less informative. This difference is easy to forget, especially since in the traditional language of the Aristotelian syllogism sentences of the form "Some As are Bs" or "Some As are not B" [$(\exists x)(Ax \wedge Bx)$ or $(\exists x)(Ax \wedge \sim Bx)$] are called "particulars", and in ordinary language 'particulars' is often synonymous with 'details' or 'specifics'. I shall try to avoid the use of the word 'particulars' altogether.

The traditional logic of the categorical syllogism is far more limited than modern predicate logic – like the minuet, in comparison to modern ball room dancing. It deals with propositions in subject-predicate form, divided into four types as follows:

A, the universal affirmative,
All As are Bs

which we would symbolize:

$$(\forall x)(Ax \rightarrow Bx)$$

E, the universal negative,
No As are Bs

which we would write either

$$(\forall x)(Ax \rightarrow \sim Bx)$$

or

$$\sim(\exists x)(Ax \wedge Bx)$$

I, the particular affirmative,
Some As are Bs

which we write:

$$(\exists x)(Ax \wedge Bx)$$

and O, the negative particular,
Some As are not B

which we write:

$$(\exists x)(Ax \wedge \sim Bx)$$

The "quantity" (universal or particular, all or some) is thought of as pertaining to the subject (*A*), and the "quality" (affirmative or negative) is thought of as pertaining to the predicate (*B*).

A surprisingly large number of ordinary-language sentences can be paraphrased so as to fit into one or another of these four patterns. In particular, the A proposition, the universal affirmative, is conveyed by a wide variety of English locutions:

>All *F*s are *G*
>All *F*s *G*
>Every *F G*s
>If something *F*s it *G*s
>Only *G*s are *F*
>Whatever *F*s *G*s
>Whoever is *F* is *G*
>Anything that *F*s is *G*
>If anything is *F* it is *G*
>Anything is *G* if it is *F*
>If one is *F* one is *G*
>etc., etc.

Singular statements, such as "Socrates is a philosopher" are traditionally assimilated to universals, as if a proper noun were no different from a common noun, except that it has only one referent. "Socrates is a philosopher," which we write

$$Ps$$

without quantifiers, is thus interpreted as "All Socrateses are philosophers" or "Everyone who is Socrates is a philosopher," or

$$(\forall x)(Sx \to Px)$$

We shall return to this issue in chapter 8.[5]

---

[5] For an insightful discussion of the relation between singular statements on the one hand and universal and particular statements on the other, in the traditional mode, see Fred Sommers, "Do We Need Identity?" *Journal of Philosophy* LXVI, 15 (Aug. 7, 1969): 499–504. Sommers makes the ingenious suggestion that a singular statement should be treated as "quantity-wild" (page 502), on the analogy of "wild" cards in poker. This analogy will become clearer in chapter 8.

Two Restricted Rules

We turn next to the restricted rules: Existential Instantiation (EI, the rule of naming, or $\exists$ elimination) and Universal Generalization (UG, or $\forall$ introduction), two rules which, in Quine's words, "cannot be directly justified, for the good reason that they serve to deduce conclusions from premises insufficient to imply them."[6]

Both rules involve the introduction and use of *pseudonames*, names used in the course of a deduction for convenience, but without external reference. A pseudoname is a name for an example, provided temporarily, so as to facilitate reasoning. In contrast, when a family names a child or when a community names a river or a town, the name attaches to an entity in the world and is intended to last a lifetime, or longer. (It does not always work out that way, but that is the intention.) A *pseudoname* is like a name in a short story; it holds its reference temporarily, for the space of an argument or a deduction.

For this purpose, as we noted earlier, we use single lower-case letters from the beginning of the alphabet; so pseudonames are immediately distinguishable from variables, but not from real names. It is helpful to be as clear as possible about the linguistic space in which they keep their identity.

Both rules must be seen as authorizing steps taken within a deduction, not as standing alone. The deductions in which they figure must be *finished* in a sense to be defined below. We shall state the two rules separately, but we shall list their restrictions together. We state first the rule of naming:

Existential Instantiation: From an existentially quantified statement we may derive an instance, provided that the pseudoname that replaces the variable of quantification is *new*.

Accordingly,

$$\frac{(\exists x)(Fx)}{\therefore Fa} \qquad \text{EI } (a)$$

---

[6] *Methods*, 2d ed., Foreword, p. vi.,

where '*a*' is new to the argument, does not appear in its conclusion, and is not a "real"name – that is, it has not been assigned to some entity in the world.

Let us imagine that I come home one day and find that someone has locked himself into the downstairs bathroom. It would be reasonable for me to call him "Joe" and use the name 'Joe' in explaining to him how to manipulate the lock so as to get out, but I am not to conclude that I know who he is or that I am talking to, say, Joe Jenkins, who lives next door. Given an existential claim, it is sensible to give a name to the item that is said to exist, so as to be able to reason about it and bring to bear upon it relevant information that I may possess, but this pseudoname is a device to be discarded before I reach a firm conclusion.

We have just seen why a pseudoname introduced by EI must be *new* in that it cannot be a real name, a name already assigned to an entity in the world or appearing in the premises or conclusion of the argument under consideration. To introduce a name "old" in this sense would give the reasoner an illusion of having knowledge she does not have. She would imagine that she knew what she was talking about. In truth she would be referring only to *some* entity known or supposed to exist – talking, that is, about "something or other."

A pseudoname introduced by EI must be new in another sense as well. It cannot have occurred already as a pseudoname within the argument in question. To introduce the same pseudoname twice would be to risk using the same expression to refer to two different things, thus inviting confusion. In ordinary life we avoid such confusion in a variety of ways, distinguishing, for example, between John Jones, Sr. and John Jones, Jr., between Grace and 'Ti-Grace, and between Belgrade, Maine, and Belgrade, Serbia. We need to use such mechanisms because there are, of course, not nearly enough names available for all the things we wish to talk about. In predicate logic we produce a similar result by *decreeing* that any one pseudoname may be introduced once only within a deduction.

Principles of Inference 159

An example of the use of Existential Instantiation is now in order. We establish the validity of a "particular" syllogism:[7]

Some burglars are caught by the police.
All burglars are dishonest.
Therefore, some who are dishonest are caught by the police.

$*\begin{cases} 1 & (\exists x)(Bx \wedge Cx) \\ 2 & (\forall x)(Bx \to Dx) \end{cases}$ PREM

To Prove: $(\exists x)(Dx \wedge Cx)$

* 3 $Ba \wedge Ca$ 1 EI (a)
* 4 Ba 3 SIMP
* 5 Ca 3 SIMP
* 6 $Ba \to Da$ 2 UI
* 7 Da 6, 4 MP
* 8 $Da \wedge Ca$ 7, 5 ADJ
* 9 $(\exists x)(Dx \wedge Cx)$ 8 EG
                         QED

Note that the premises do not warrant the "conclusion" shown at step 8: Albert is dishonest and was caught by the police. This statement is unintelligible in isolation because we do not know who Albert is – or unwarranted if we imagine that we do know who Albert is; we don't. The premises do warrant the conclusion shown at step 9, in which 'a' does not appear.

Notice also the letter 'a' shown in parenthesis to the right of step 3. This is a *flag*[8] – for our purposes a red flag, saying "watch out!" It alerts the arguer to the necessity for checking on whether the restrictions on EI or UG are being violated. We flag a letter when it is introduced by EI or eliminated by UG. The restrictions on EI and UG will be articulated in terms of flagged letters.

---

[7] A "particular" syllogism, besides a universal premise (needed in every case), has one "particular" (existential) premise, and a "particular" (existential) conclusion.
[8] *Methods*, 1st ed., p. 161; 2d ed., p. 160.

Predicate Logic

Universal Generalization: From a statement containing a pseudoname we may derive a universally quantified statement of which it is an instance, provided that the pseudoname, which is being replaced by the variable of quantification, has been introduced without special conditions; the item it refers to is not a *special* case.

Accordingly,

$$\frac{Fa}{\therefore (\forall x)(Fx)} \quad \text{UG } (a)$$

where there are no special conditions on a and 'a' is not a "real" name.

Generalizing in informal reasoning, as we all know, is a delicate matter. It is illegitimate to generalize from a special case. If Susie, who is young and strong and a seasoned hiker, found it easy to climb Mt. Katahdin, it does not follow that everyone will find it easy to climb Mt. Katahdin. On the other hand, it is legitimate (and often useful) to generalize from a single case when we are certain that our reasons for believing what holds in that single case would be just as telling in any other case. But we must be certain that this is so, that we are not dealing with a special case.

The restrictions on the rules EI and UG, listed below, are designed to prevent us from using UG to generalize inappropriately, as well as to prevent us from using EI inappropriately, creating confusion by giving the same name to different things. These restrictions make clear what is intended by the requirement that the pseudoname introduced by EI be *new* and by the parallel requirement that the pseudoname eliminated by UG not be *special*.

The standard examples of the correct use of UG come from geometry. We prove, for example, that if the triangle ABC is equilateral, the triangle ABC is also equiangular; it follows that all equilateral triangles are equiangular, since the triangle ABC is *any* triangle.[9] Here we shall use instead an example from logic,

---

[9] One must be very careful. For carelessness can lead to fallacy. A famous argument, familiar to geometers, begins like this:

and establish the validity of a "universal" syllogism, a classical AAA syllogism in the first figure:

All winged things can fly.
All birds have wings.
Therefore, all birds can fly.

Take any triangle ABC. Drop a perpendicular from the vertex A to the line through B and C, and mark the point of intersection D.

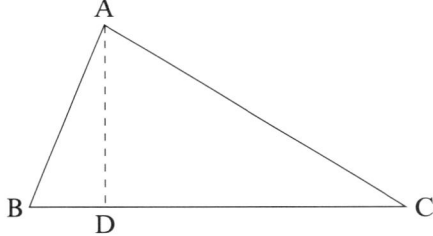

We have, then, BC = BD + DC.
From this innocent beginning one can prove that 2 = 0.
The problem is that we do *not* have BC = BD + DC; that equation was read from a diagram of a special case. If the angle B had been obtuse instead of acute, that is, we would have got BC = DC − DB, and, if B were a right angle, BC would equal DC. The illusion of generality produced by careless (or ingenious but malevolent) diagramming can be misleading.

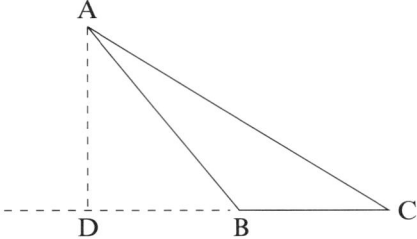

For a related geometric "proof" that every triangle is isosceles, see W.E. Johnson, *Logic*, Part II (Cambridge: University Press, 1922; New York: Dover, 1964), pp. 206/7. This fallacious "proof" is presented and explained by Johnson in the context of an illuminating discussion of generalization in geometry.

$*\begin{cases} 1 & (\forall x)(Wx \to Fx) \\ 2 & (\forall x)(Bx \to Wx) \end{cases}$ PREM

To Prove: $(\forall x)(Bx \to Fx)$

*	3	$Wa \to Fa$	1 UI
*	4	$Ba \to Wa$	2 UI
**	5	Ba	PREM (for CP)
**	6	Wa	4, 5 MP
**	7	Fa	3, 6 MP
*	8	$Ba \to Fa$	5–7 CP
*	9	$(\forall x)(Bx \to Fx)$	8 UG (a)
			QED

It should be noted that this deduction could be done more efficiently, after step 4, like this:

*	5'	$Ba \to Fa$	4, 3 CHAIN
*	6'	$(\forall x)(Bx \to Fx)$	5' UG (a)
			QED

But it could *not* be done like this:

* 3″ $(\forall x)(Bx \to Fx)$ 2, 1 CHAIN

For the CHAIN rule authorizes the deduction of a conditional statement from two other conditional statements, and these premises and this conclusion are *not* conditionals. The pair of principles UI and UG allow us to move from universal statements to singular statements (by UI), do some work in propositional logic (here, steps 3 to 8 or steps 3 to 5'), and then move back again (by UG). This process may not be bypassed.

The following argument, which is quite different from the argument above, *is* an instance of the CHAIN rule:

If everything has wings, everything can fly.
If everything is a bird, everything has wings.
Therefore, if everything is a bird, everything can fly.

$(\forall x)(Wx) \to (\forall x)(Fx)$
$(\forall x)(Bx) \to (\forall x)(Wx)$
∴ $(\forall x)(Bx) \to (\forall x)(Fx)$ CHAIN
                                              QED

I now list the restrictions on EI and UG, which should be thought of as part of what is meant by these two principles. It would be well for the reader to check the two proofs I have given – the one about the burglars and the one about the birds – to see that they do not violate these restrictions.

1. A deduction in which any flag occurs must be *finished*; that is, no flagged letter may occur either in its conclusion or in any of its undischarged premises.

2. Within a deduction, no letter may be flagged twice. Using the same flag twice is the *fallacy of double flagging*.

Eventually we shall need further restrictions, enjoining against cross flagging and cyclical flagging. But these are irrelevant until we come to the logic of relations; so I omit them here.[10]

These restrictions spell out what is intended by requiring that the pseudoname introduced by EI must be "new" and that the pseudoname eliminated in UG may not be "special". I cite next a number of obviously fallacious deductions that contravene them, in order to make clear why the restrictions are needed.[11]

A. Someone is in the closet and wants to get out.
   Therefore, Joe Jenkins is in the closet.

* 1 $(\exists x)(Cx \land Wx)$ PREM
* 2 $Cj \land Wj$ 1 EI (j)
* 3 Cj 2 SIMP

This is an *unfinished* deduction, with a flagged letter, j, occurring in its conclusion. This conclusion does not follow from the prem-

---
[10] See Part III, chapter 6.
[11] Note that we omit the "QED" at the end of a fallacious deduction; such a deduction is not successful and should not be emulated.

ise. But if we continue with a further step that eliminates the flagged letter:

* 4 $(\exists x)(Cx)$ 3 EG
(Somebody is in the closet)

QED

we arrive at a valid conclusion.

  B. If Mary eats strawberries, Mary gets a rash.
     Therefore, everyone who eats strawberries gets a rash.

* 1 $Sm \to Rm$ PREM
* 2 $(\forall x)(Sx \to Rx)$ 1 UG (m)

This is also an *unfinished* deduction (the flagged letter 'm' occurs in its premise, which is not discharged), but this deduction is not so easily repaired. For generalizing in this case to be acceptable we would have to provide a rationale for the premise that would make clear that Mary was not a special case. This rationale is missing.

By way of comparison with this invalid argument, let us look back at the valid syllogistic argument about the birds (p. 162). UG is invoked at step 9, where "All birds can fly" is derived from "If Albert is a bird, Albert can fly" (step 8). But this move is legitimate, because Albert is not a special case. Step 8 has been derived from two universal premises, which apply to everything, not just to Albert. The flagged letter 'a' does not appear in premises 1 and 2; 'a' does appear in premise 5, but premise 5 has been discharged. This argument is in sharp contrast with the fallacious argument about Mary and the strawberries.

  C. Some animals are cats and some animals are dogs.
     Therefore, some cats are dogs.

* 1 $(\exists x)(Ax \land Cx) \land (\exists x)(Ax \land Dx)$ PREM

To Prove: $(\exists x)(Cx \land Dx)$

* 2 $(\exists x)(Ax \land Cx)$ 1 SIMP
* 3 $(\exists x)(Ax \land Dx)$ 1 SIMP
* 4 $Aa \land Ca$ 2 EI (a)

* 5	$Aa \wedge Da$	3 EI (*a*)
* 6	Ca	4 SIMP
* 7	Da	5 SIMP
* 8	$Ca \wedge Da$	4, 5 ADJ
* 9	$(\exists x)(Cx \wedge Dx)$	9 EG

The fallacy of double flagging occurs at step 5, and the fallacious conclusion could not be drawn without it. Step 5 could be corrected to

* 5'	$Ab \wedge Db$	3 EI (*b*)

but that step would be pointless. $Ca \wedge Db$ at step 8 would not yield the conclusion we were interested in.

D. Something is interesting.
   Therefore, everything is interesting.

* 1	$(\exists x)(Ix)$	PREM
* 2	Ia	1 EI (*a*)
* 3	$(\forall x)(Ix)$	2 UG (*a*)

Argument D is obviously invalid. Technically, it commits the fallacy of double flagging. It is perhaps worth noting that the converse argument, which involves no flags, is just as obviously *valid*:

D'. Everything is interesting.
   Therefore, something is interesting.

* 1	$(\forall x)(Ix)$	PREM
* 2	Ia	1 UI
* 3	$(\exists x)(Ix)$	2 EG
		QED

The validity of this argument reflects the assumption, noted earlier, that we are working in a nonempty universe. At step 2 we use UI to introduce the pseudoname *a*, without a flag. There is no fallacy. At step 3 we could use UG to draw the conclusion $(\forall x)(Ix)$ or $(\forall y)(Iy)$, instead of $(\exists x)(Ix)$, again without fallacy, but that would be uninteresting – a mere repetition or reformulation of the premise.

E. Everything is interesting or uninteresting. Therefore, everything is interesting or everything is uninteresting.

```
  * 1  (∀x)(Ix ∨ ~Ix)                PREM
  * 2  Ia ∨ ~Ia                      1 UI
L * 3  Ia                            PREM (for DIL)
L * 4  (∀x)(Ix)                      3 UG (a)
L * 5  (∀x)(Ix) ∨ (∀x)(~Ix)          4 THIN
R * 6  ~Ia                           PREM (for DIL)
R * 7  (∀x)(~Ix)                     6 UG (a)
R * 8  (∀x)(Ix) ∨ (∀x)(~Ix)          7 THIN
  * 9  (∀x)(Ix) ∨ (∀x)(~Ix)          2, 3-5, 6-8 DILemma
```

The fallacy of double flagging occurs at step 7, since a has been flagged already at step 4. This fallacy is essential to the "proof", which is obviously fallacious, since it moves from a premise that is known to be true to a conclusion that is known to be false.

The "deductions" for the three fallacious arguments C, D, and E, shown above, all commit the fallacy of double flagging. The two flags in the "deduction" for C result from two uses of EI, the rule of naming: two entities are assigned the same pseudoname, inviting confusion. The two flags in the "deduction" for D result from one use of EI, followed by a use of UG: we have generalized from a flagged pseudoname – a special case. In the "deduction" for E the two flags result from two uses of UG – again, the second flagging alerts the reader to generalization from a special case. The reader will do well to reread all five of the prohibited deductions and make sure that the fallaciousness of all three arguments is intuitively clear: their conclusions do *not* follow from their premises.

Arguments C and E make clear that the following (fallacious but tempting) "implications" do not hold:

$$[(\exists x)(Fx) \land (\exists x)(Gx)] \to (\exists x)(Fx \land Gx)$$
$$(\forall x)(Fx \lor Gx) \to [(\forall x)(Fx) \lor (\forall x)(Gx)]$$

Therefore, $(\exists x)(Fx) \land (\exists x)(Gx)$ is *not* equivalent to $(\exists x)(Fx \land Gx)$, and $(\forall x)(Fx) \lor (\forall x)(Gx)$ is *not* equivalent to $(\forall x)(Fx \lor Gx)$, and so these pairs of formulas are not interchangeable. This fact should alert the reader to the importance of careful attention to the scope of quantifiers. Mistaking a conjunction of narrow-scope existentials for a wide-scope existential with a conjunctive open sentence or mistaking a disjunction of narrow-scope universals for the "corresponding" wide-scope universal with disjunctive open sentence can make the difference between confusion and clarity of reasoning.

The following two equivalences, on the other hand, do hold:

$$[(\exists x)(Fx) \vee (\exists x)(Gx)] \leftrightarrow (\exists x)(Fx \vee Gx)$$
$$[(\forall x)(Fx) \wedge (\forall x)(Gx)] \leftrightarrow (\forall x)(Fx \wedge Gx)$$

and are often useful. The reader will do well to carry out the proofs. In doing so one must be careful to avoid the fallacy of double flagging.

This completes our exposition of the four basic principles of inference needed for monadic predicate logic (in addition to the propositional rules of inference introduced in chapter 1), i.e., the unrestricted rules UI and EG and the restricted rules EI and UG. Before we go on to discuss some auxiliary rules, let me remind the reader of two quantificational principles of interchange, mentioned in passing at the beginning of this chapter. These make explicit the ways in which interchange occurs in quantificational contexts.

Quantificational INTerchange (Q-INT): Any two open sentences whose instances have been shown to be equivalent, or deducible one from the other, by the methods of truth-functional logic, are interchangeable.

Change of Variable (CV): Any two quantified statements alike except for having different variables of quantification are equivalent and, accordingly, interchangeable.

Quantificational Negation

We turn now to the concept of negation in quantification theory. It should be clear from the outset that the negation of a universally quantified statement will be an existential. To deny, for example, that everything is material is to affirm that *something* is not, that is, that something is immaterial; it is *not* to affirm that *everything* is immaterial. Similarly, the negation of an existentially quantified statement will be a universal. To deny that something (anything) is material is to affirm that everything is immaterial. This observation yields a principle of Quantifier Negation (QN) and a pair of quantificational equivalences analogous to the NAND and NOR of truth-function theory.

Quantifier Negation: The negation of a universally quantified statement is the existentially quantified negation of the open sentence of the first. The negation of an existentially quantified statement is the universally quantified negation of the open sentence of the first.

Accordingly,

$$QN: \sim(\forall x)(Fx) \leftrightarrow (\exists x)(\sim Fx)$$
$$\sim(\exists x)(Fx) \leftrightarrow (\forall x)(\sim Fx)$$

These equivalences are not independently postulated or stipulated; they follow from principles already stated. Let us carry out the proofs. We start with the second one:

To prove: $\sim(\exists x)(Fx) \leftrightarrow (\forall x)(\sim Fx)$

*	1 $(\forall x)(\sim Fx)$	PREM (for CP, right to left)
**	2 $(\exists x)(Fx)$	PREM (for RED)
**	3 Fa	2 EI (a)
**	4 $\sim Fa$	1 UI
**	5 $Fa \land \sim Fa$	3, 4 ADJ
*	6 $\sim(\exists x)(Fx)$	2-5 RED
	7 $(\forall x)(\sim Fx) \rightarrow \sim(\exists x)(Fx)$	1-6 CP

In going from right to left (as above) the strategy of reductio comes immediately to mind and can be used promptly (see step 2), since the proposed conclusion (see step 6) is a negation.

Going from left to right, aiming to show, that is, that

$$\sim(\exists x)(Fx) \rightarrow (\forall x)(\sim Fx)$$

reductio will be used again, but a more devious strategy will be needed. Since what is to be shown is $(\forall x)(\sim Fx)$, the last step will have to be universal generalization; so we must aim to show an *instance* of $(\forall x)(\sim Fx)$ on the basis of which that last step can be taken. We cannot choose to prove $\sim Fa$, since that would require us to flag a in deriving $(\forall x)(\sim Fx)$, and a has already been flagged at step 3. We choose instead to prove $\sim Fb$. So the second step in this part of the deduction is the auxiliary premise Fb (see step 9 below).

```
 *   8  ~(∃x) Fx                              PREM (for CP)
**   9  Fb                                    PREM (for RED)
**  10  (∃x)(Fx)                              9 EG
**  11  (∃x)(Fx) ∧ ~(∃x)(Fx)                  10, 8 ADJ
 *  12  ~Fb                                   9–11 RED
 *  13  (∀x)(~Fx)                             12 UG (b)
    14  ~(∃x)(Fx) → (∀x)(~Fx)                 8–13 CP
    15  ~(∃x)(Fx) ↔ (∀x)(~Fx)                 7,14 ADJ, DEF ↔
                                              QED
```

This equivalence, like NOR, is built into our ordinary use of language. If one is asked to paraphrase, say, "Nothing is certain" with an eye to restating it in symbols, one is equally likely to think of this as a universal claim:

> Everything is uncertain.
> $(\forall x)(\sim Cx)$

or to think of it as a denial of existence:

> There is nothing that is certain.
> $\sim(\exists x)(Cx)$

This goes for such locutions as 'none', 'nobody', 'never', as well. The other equivalence is not quite so obvious.

To prove: $\sim(\forall x)(Fx) \leftrightarrow (\exists x)(\sim Fx)$ – starting from right to left:

```
 *   1  (∃x)(~Fx)              PREM (for CP)
**   2  (∀x)(Fx)               PREM (for RED)
**   3  ~Fa                    1 EI (a)
**   4  Fa                     2 UI
**   5  Fa ∧ ~Fa               3, 4 ADJ
 *   6  ~(∀x)(Fx)              2–5 RED
     7  (∃x)(~Fx) → ~(∀x)(Fx)  1–6 CP
```

```
 *  8  ~(∀x)(Fx)                      PREM (for CP)
**  9  Fb                             PREM (for RED)
** 10  (∀x)(Fx)                       9 UG (b)
** 11  (∀x)(Fx) ∧ ~(∀x)(Fx)           10, 8 ADJ
 * 12  ~Fb                            9-11 RED
 * 13  (∃x)(~Fx)                      12 EG
   14  ~(∀x)(Fx) → (∃x)(~Fx)         8-13 CP
   15  ~(∀x)(Fx) ↔ (∃x)(~Fx)         7, 14 ADJ, DEF ↔
                                      QED
```

Another (doubly redundant) pair of equivalences is worth mentioning. These are of historical importance because they define the "square of opposition" of Aristotelian logic. They are also very convenient.

QN': The negation of a universal affirmative proposition is the corresponding negative particular. The negation of an affirmative particular is the corresponding negative universal.

Accordingly,

$$\text{QN'}: \quad \sim(\forall x)(Ax \to Bx) \leftrightarrow (\exists x)(Ax \land \sim Bx)$$
$$\sim(\exists x)(Ax \land Bx) \leftrightarrow (\forall x)(Ax \to \sim Bx)$$

Since the four principles of quantification theory: UI, EG, EI, and UG, move from general statements to their instances, or vice versa, but give no direct indication of what follows from denials of universality or denials of existence, QN (and QN') will often be useful for *internalizing* negation and so making it possible to reason with such denials. I cite two examples.

F. All students take algebra.
   Not all students take music.
   Therefore, some who take algebra do not take music.

```
  ⎧ 1  (∀x)(Sx → Ax)
* ⎨                                   PREM
  ⎩ 2  ~(∀x)(Sx → Mx)
```

Principles of Inference

To Prove: $(\exists x)(Ax \wedge \sim Mx)$

*	3 $(\exists x)(Sx \wedge \sim Mx)$	2 QN'
*	4 $Sa \wedge \sim Ma$	3 EI (*a*)
*	5 Sa	4 SIMP
*	6 $\sim Ma$	4 SIMP
*	7 $Sa \to Aa$	1 UI
*	8 Aa	7, 5 MP
*	9 $Aa \wedge \sim Ma$	8, 6 ADJ
*	10 $(\exists x)(Ax \wedge \sim Mx)$	9 EG
		QED

The temptation to use, instead of 3 and 4, a different "step" 3':

*	3' $\sim(Sa \to Ma)$	2 UI

should be resisted, even though 3' looks equivalent to 4 (it is not, because 4 carries a flag: *a*). UI makes sense with 1 (see step 7), but not with 2, which is *not* a universal. Because the fallacious "step" 3' carries no flag, it would tempt us to derive a variety of universal conclusions by UG from any of steps 5, 6, 8, 9, none of which would make sense. (Everyone is a student, from 5; No one takes music, from 6; Everyone takes algebra, from 8; Everyone takes algebra but not music, from 9.)

G. No men are immortal.
No angels are mortal.
Therefore, no men are angels.

* {	1 $\sim(\exists x)(Hx \wedge \sim Mx)$	PREM
	2 $\sim(\exists x)(Ax \wedge Mx)$	

To Prove: $\sim(\exists x)(Hx \wedge Ax)$

*	3 $(\forall x)(Hx \to Mx)$	1 QN'
*	4 $(\forall x)(Ax \to \sim Mx)$	2 QN'

Negation must be internalized at least in the premises, but there is choice about the conclusion. Either of the following will do:

* 5	$Ha \to Ma$	3 UI
* 6	$Aa \to {\sim}Ma$	4 UI
* 7	$Ma \to {\sim}Aa$	6 INT(Contraposition)
* 8	$Ha \to {\sim}Aa$	5, 7 CHAIN
* 9	$(\forall x)(Hx \to {\sim}Ax)$	8 UG (a)
		QED

or

** 5'	$(\exists x)(Hx \wedge Ax)$	PREM (for RED)
** 6'	$Ha \wedge Aa$	5' EI (a)
** 7'	Ha	6' SIMP
** 8'	Aa	6' SIMP
** 9'	$Ha \to Ma$	3 UI
** 10'	$Aa \to {\sim}Ma$	4 UI
** 11'	Ma	9', 7' MP
** 12'	${\sim}Ma$	10', 8' MP
** 13'	$Ma \wedge {\sim}Ma$	11', 12' ADJ
* 14'	${\sim}(\exists x)(Hx \wedge Ax)$	5'–13' RED
		QED

Either way, there will be one flag: where a pseudoname is eliminated, by UG – see step 9 – or where one is introduced by EI – see step 6'. Negation must be internalized wherever UI or EI is to be used (see steps 3 and 4, in preparation for steps 5 and 6 or for steps 9' and 10').

The Rules of Passage

Deductions such as these have required careful attention to the scope of quantifiers. The reader will remember that the scope of a quantifier is the open sentence associated with it, marked with parentheses or brackets; it defines the linguistic space within which the variable of quantification retains its identity and so achieves cross reference. When the scopes of all the quantifiers in a formula are narrow, the formula is said to be in *pure* form. When the scopes are wide and the quantifiers all precede a single open sentence, or *nexus*, the formula is in *prenex* form. Equivalences such as QN and QN′ allow us to shrink or stretch the scope of certain quantifiers. In the negations (the left-hand sides of the four equivalences) the scope is narrow, with the negation sign outside it; on the right-hand sides negation has been internalized, and so the scope is wide, including the negation sign.

I shall cite next a list of eight equivalences – "rules of passage"[12] – that warrant shrinking or stretching the scopes of quantifiers in eight other cases. Like the auxiliary principles of chapter 1 (modus tollens, etc.), the rules of passage are all provable and therefore redundant; they are useful as short cuts, and also occasionally helpful for understanding the structure of our reasoning.

But, before we deal with the rules of passage, let us consider an example of the problem that these rules address.

If anyone telephoned, Henry took a message.
Joe telephoned.
So Henry took a message.

This little argument is patently valid. How should it be formalized? The first premise can be read in either of two ways: first, as a *conditional* with antecedent "anyone telephoned," or "someone or other telephoned," and consequent "Henry took a message":

$$(\exists x)(Tx) \to Mh$$

and second as a *universal*, where "Henry took a message" is the consequent of the *open sentence*, that is,

$$(\forall x)(Tx \to Mh)$$

The deductions will be different, depending on which version of the first premise we use. In this example, both deductions are short and easy to find, but in more complicated cases there may be more significant differences either in ease of reading or in convenience of work. Here are the deductions:

---

[12] *Methods*, 4th ed., pp. 142–143.

```
*  ⎧ 1  (∃x)(Tx) → Mh           PREM
   ⎩ 2  Tj
*    3  (∃x)(Tx)                2 EG
*    4  Mh                      1, 3 MP
                                QED

*  ⎧ 1'  (∀x)(Tx → Mh)          PREM
   ⎩ 2'  Tj
*    3'  Tj → Mh                1' UI
*    4'  Mh                     3', 2' MP
                                QED
```

The fact that these two deductions are both successful (and both use only the unrestricted rules) suggests that the two readings of the first premise are equivalent, as indeed they are; see Rule of Passage 7, below. We can, accordingly, choose as we see fit between the two readings of the first premise.

In another argument, however, an argument that might strike a careless reader as similar to this one, we have no such choice.

If anyone telephoned, he (or she) left a message.
George telephoned.
Therefore, George left a message.

The structural difference between these arguments is that in this one the subject of the "consequent" of the first premise refers back to the subject of the "antecedent." A formalization such as "$(\exists x)(Tx) \to Mx$" would fail, because the third 'x', the 'x' of "x left a message," would "dangle"; it cannot refer back to the "x" who telephoned (although it is intended to so refer), because the quantifier does not cover it. So in this case the narrow-scope alternative in formalizing the first premise is *not* available. The argument must be formalized like this, with a wide-scope universal quantifier covering the whole first premise:

$$(\forall x)(Tx \to Mx)$$
$$\underline{Tg}$$
$$\therefore\ Mg$$

There are a number of morals to be drawn from consideration of these arguments. First, in formalizing sentences which sound like conditionals (or other types of compound sentence) and which involve generality, it is wise to be sensitive to questions of cross reference. If any cross reference is intended between items that occur in what will be formalized as different open sentences, one must be careful to make sure that the quantifiers in the symbolic formulation have scope wide enough to convey that cross reference. Sensitivity to questions of cross reference is valuable, not only for proper formalization, but also for knowing what one is talking about.

Second, if there is no cross reference intended between clauses of a sentence to be formalized, that sentence may be rendered either with narrow-scope quantifiers, which are relatively easy to read, or with wide-scope quantifiers, which may be more convenient. The equivalences that follow make clear why these alternatives are available.

It will be useful not only to understand that these equivalences hold, but also to be sensitive to the differences in meaning between the left-hand sides (where the quantifiers have narrow scope), and the right-hand sides (where the scope is wide). I will comment on a few of them and do deductions for the last two; the reader will do well to pay attention to them all. 'Fx' throughout stands for any open sentence in x, and 'p' throughout is any sentence that does *not* contain 'x'.

(1) $[p \land (\forall x)(Fx)] \leftrightarrow (\forall x)(p \land Fx)$

(2) $[p \land (\exists x)(Fx)] \leftrightarrow (\exists x)(p \land Fx)$

(3) $[p \lor (\forall x)(Fx)] \leftrightarrow (\forall x)(p \lor Fx)$

(4) $[p \lor (\exists x)(Fx)] \leftrightarrow (\exists x)(p \lor Fx)$

(5) $[p \to (\forall x)(Fx)] \leftrightarrow (\forall x)(p \to Fx)$

(6) $[p \to (\exists x)(Fx)] \leftrightarrow (\exists x)(p \to Fx)$

(7) $[(\exists x)(Fx) \to p] \leftrightarrow (\forall x)(Fx \to p)$

(8) $[(\forall x)(Fx) \to p] \leftrightarrow (\exists x)(Fx \to p)$

I cite also eight truth-functional equivalences, analogues of the eight rules of passage, which may be helpful for understanding those rules, just as NAND and NOR were helpful for understanding QN. Two of these equivalences, (1') and (4'), are trivial; for their two sides are exactly alike, except for unnecessary parentheses and repetitious p on the right-hand side. The other six are instances of distribution rules; see the list of equivalences for chapter 1. (8) and (8'), alike, are counterintuitive, but true.

(1') $[p \wedge (Fa \wedge Fb)] \leftrightarrow [(p \wedge Fa) \wedge (p \wedge Fb)]$

(2') $[p \wedge (Fa \vee Fb)] \leftrightarrow [(p \wedge Fa) \vee (p \wedge Fb)]$

(3') $[p \vee (Fa \wedge Fb)] \leftrightarrow [(p \vee Fa) \wedge (p \vee Fb)]$

(4') $[p \vee (Fa \vee Fb)] \leftrightarrow [(p \vee Fa) \vee (p \vee Fb)]$

(5') $[p \rightarrow (Fa \wedge Fb)] \leftrightarrow [(p \rightarrow Fa) \wedge (p \rightarrow Fb)]$

(6') $[p \rightarrow (Fa \vee Fb)] \leftrightarrow [(p \rightarrow Fa) \vee (p \rightarrow Fb)]$

(7') $[(Fa \vee Fb) \rightarrow p] \leftrightarrow [(Fa \rightarrow p) \wedge (Fb \rightarrow p)]$

(8') $[(Fa \wedge Fb) \rightarrow p] \leftrightarrow [(Fa \rightarrow p) \vee (Fb \rightarrow p)]$

Rule of passage (1) tells us, for example, that "The sun is shining and everyone is in good humor" is equivalent to "Everyone is such that the sun is shining and he is in good humor"; (4) tells us that "Either the roof will be fixed or someone will get wet" is equivalent to "Someone is such that either the roof will be fixed or he will get wet." The eight expressions involving "if" are a little easier to distinguish. The word 'certain' suggests wide scope of the existential quantifier[13]; the word 'any' suggests wide scope of the universal.[14] So (6) might tell us that "If there is a storm, someone will be hurt" is equivalent to "Certain people will be hurt if there is a storm", and (7) that "If someone pushes the button the elevator will come up" is equivalent to "If anyone pushes the button, the elevator will come up". Here is a deduction for (7):

---

[13] This was suggested to me by a student, Charles Johnson, twenty years ago.
[14] *Methods*, 4th ed, p. 144.

To Prove: $[(\exists x)(Fx) \to p] \leftrightarrow (\forall x)(Fx \to p)$

*	1	$(\exists x)(Fx) \to p$	PREM (for CP, left to right)
**	2	Fa	PREM (for CP)
**	3	$(\exists x)(Fx)$	2 EG
**	4	p	1, 3 MP
*	5	$Fa \to p$	2–4 CP
*	6	$(\forall x)(Fx \to p)$	5 UG (a)
	7	$[(\exists x)(Fx) \to p] \to (\forall x)(Fx \to p)$	1–6 CP
*	8	$(\forall x)(Fx \to p)$	PREM (for CP, right to left)
**	9	$(\exists x)(Fx)$	PREM (for CP)
**	10	Fb	9 EI (b)
**	11	$Fb \to p$	8 UI
**	12	p	11, 10 MP
*	13	$(\exists x)(Fx) \to p$	9–12 CP
	14	$(\forall x)(Fx \to p) \to [(\exists x)(Fx) \to p]$	8–13 CP
	15	$[(\exists x)(Fx) \to p] \leftrightarrow (\forall x)(Fx \to p)$	7, 14 ADJ, DEF $\leftrightarrow$ QED

Rules (1) to (6) are straightforward. (7), although it involves a change of quantifier, seems reasonable. (8), like (8′) however, may be counterintuitive: "If everything is F, then p" does not seem equivalent to "If certain things are F then p". "If both Fa and Fb then p" does not seem equivalent to "Either if Fa then p or if Fb then p". The problem in each case is with the implication from left to right, not from right to left. It is obvious that if there is something such that its being F is sufficient for p, everything's being F will also be sufficient for p; likewise, that if Fa is sufficient for p or Fb is sufficient for p then the occurrence of both Fa and Fb will be sufficient for p. But the converses are troublesome. The contrast between a pair of examples may be helpful.

First, let $Fx = x$ jumps up and down on the roof, and let $p =$ the roof caves in. "If everyone jumps up and down on the roof, the roof caves in" does not seem to imply that if certain people jump up and down on the roof, the roof will cave in or that there is at least one person such that if she jumps up and down on the roof, the roof will cave in. Nor does "If Albert and Bella both jump up and down on the roof, the roof will cave in" seem to imply that "If Albert jumps up and down on the roof the roof will cave in or if Bella jumps up and down on the roof the roof will cave in."

Second, let $Fx = x$ tries to solve this algebra problem, and let p = the problem is solved. "If all of us try to solve the algebra problem, the problem will be solved" does seem to imply that there is at least one among us such that, if she tries, the problem will be solved. Likewise, "If Albert and Bella both try to solve the algebra problem, it will be solved" does seem to imply that if Albert tries it will be solved or if Bella tries it will be solved.

The difference between the two cases is that we tend to imagine everyone jumping up and down on the roof together, but everyone trying to solve an algebra problem separately. The togetherness suggested in the roof example is misleading. '$(\forall x)(Fx)$' does not suggest togetherness or reinforcement any more than does the logical "and", or '$\wedge$'. If, in the roof example, we imagine people jumping up and down on the roof on different days, the equivalence will appear more reasonable.

There is another way of looking at (8) and (8') which may also help to make them seem acceptable; we look at the claims that the negations of the formulas are equivalent.

$\sim[(\forall x)(Fx) \rightarrow p] \leftrightarrow \sim(\exists x)(Fx \rightarrow p)$
$\sim[(Fa \wedge Fb) \rightarrow p] \leftrightarrow \sim[(Fa \wedge p) \vee (Fb \wedge p)]$

Since, as we saw in chapter 2, '$\sim p \leftrightarrow \sim q$' is equivalent to '$p \leftrightarrow q$', acceptance and understanding of the above equivalences may make (8) and (8') seem more reasonable. If the denials of the statements are seen to be equivalent, the statements themselves must be equivalent, too. But $\sim[(\forall x)(Fx) \rightarrow p]$ tells us, by NIF, that $(\forall x)(Fx) \wedge \sim p$, i.e., that everyone jumps up and down but the roof doesn't fall in. And $\sim(\exists x)(Fx \rightarrow p)$ tells us, by QN, that $(\forall x) \sim(Fx \rightarrow p)$, and by NIF, that $(\forall x)(Fx \wedge \sim p)$, i.e. that everyone is such that he jumps up and down but the roof doesn't fall in.

$[(\forall x)(F x) \wedge \sim p] \leftrightarrow (\forall x)(Fx \wedge \sim p)$

is just an instance of (1), which is not counterintuitive. Likewise, to deny that $(Fa \wedge Fb) \rightarrow p$ is to claim that $Fa \wedge Fb \wedge \sim p$; and to deny that $(Fa \to p) \lor (Fb \to p)$ is to claim that $Fa \land {\sim} p \land Fb \land {\sim} p$, which comes, repetitiously, to the same thing.

Here is a deduction for (8) – starting from left to right, with the counterintuitive implication:

To Prove: $[(\forall x)(Fx) \to p] \leftrightarrow (\exists x)(Fx \to p)$

*	1	$(\forall x)(Fx) \to p$	PREM (for CP, left to right)
**	2	Fa	PREM (for CP)
**	3	$(\forall x)(Fx)$	2 UG (*a*)
**	4	p	1, 3 MP
*	5	$Fa \to p$	2–4 CP
*	6	$(\exists x)(Fx \to p)$	5 EG

[Note that $(\forall x)(Fx \to p)$ by UG (*a*) would be fallacious at step 6, for that move would commit the fallacy of double flagging.]

	7	$[(\forall x)(Fx) \to p] \to (\exists x)(Fx \to p)$	1–6 CP
*	8	$(\exists x)(Fx \to p)$	PREM (for CP, right to left)
*	9	$Fb \to p$	8 EI (*b*)
**	10	$(\forall x)(Fx)$	PREM (for CP)
**	11	Fb	10 UI
**	12	p	9,11 MP
*	13	$(\forall x)(Fx) \to p$	10–12 CP
	14	$(\exists x)(Fx \to p) \to [(\forall x)(Fx) \to p]$	8–13 CP
	15	$[(\forall x)(Fx) \to p] \leftrightarrow (\exists x)(Fx \to p)$	7, 14 ADJ, DEF $\leftrightarrow$ QED

Strategy in Deduction

Before going on to deal with tests for the validity of arguments in predicate logic, it may be well to review some of the deductive strategies exhibited in the deductions presented in this chapter.

First, when the desired conclusion is a quantified statement, plan to deduce an instance of that statement as a last step before obtaining the desired conclusion itself. Notice right away (so as to be able to choose supplementary premises wisely) whether the final conclusion will have to be derived from the instance by UG, which requires flagging, or by EG, which does not.

Second, switch pseudonames if possible where this would be helpful to avoid double flagging. If it is not possible, as in the fallacious proof of argument E (Everything is interesting or uninteresting; therefore, everything is interesting or everything is uninteresting), that may be a signal that one is trying to do something that cannot be done.

Third, always, if possible, use EI before UI. Using EI late in a deduction may force the introduction of unnecessary pseudonames and muddy the waters.

Finally, recall the strategies learned in Part I. To prove a conditional, take its antecedent as premise and try to deduce its consequent. To prove a negation, assume the statement negated and try to deduce a contradiction. To use a disjunction, assume its disjuncts separately and try to deduce a joint conclusion, or, alternatively, use disjunctive elimination. When in doubt, try reductio. And remember always that there may be many ways to carry out a deduction. If one person does it one way and another person does it another way, they may both be right.

Reminders for Chapter 4

Rules of Inference for Predicate Logic

Two Unrestricted Rules

Universal Instantiation: From a universally quantified statement any of its instances, without restriction, follows.

Accordingly,

$$\frac{(\forall x)(Fx)}{\therefore Fa} \qquad \text{UI}$$

Existential Generalization: An existentially quantified statement is deducible from any of its instances, without restriction.

Accordingly,

$$\frac{Fa}{\therefore (\exists x)(Fx)} \quad \text{EG}$$

Two Restricted Rules

Existential Instantiation: From an existentially quantified statement we may derive an instance, provided that the pseudoname that replaces the variable of quantification is *new*.

Accordingly,

$$\frac{(\exists x)(Fx)}{\therefore Fa} \quad \text{EI } (a)$$

Universal Generalization: From a statement containing a pseudoname we may derive a universally quantified statement of which it is an instance, provided that the pseudoname, which is replaced by the variable of quantification, does *not* refer to a *special* case.

Accordingly,

$$\frac{Fa}{\therefore (\forall x)(Fx)} \quad \text{UG } (a)$$

The restrictions, which spell out what is meant by the requirements that the pseudoname for EI be *new* and that what is referred to by the pseudoname in UG *not* be *special*, are as follows:

1. A deduction in which any flag occurs must be *finished*; that is, no flagged letter may occur either in its conclusion or in any of its undischarged premises.

2. Within a deduction no letter may be flagged twice. Using the same flag twice is the *fallacy of double flagging.*

(A third restriction is postponed until chapter 6.)

INTerchange: Any two statements that have been shown to be equivalent, or deducible one from the other, are interchangeable.

Q-INTerchange: Any two open sentences whose instances have been shown to be truth-functionally equivalent are interchangeable in quantificational contexts.

Change of Variable (CV): Any two quantified statements alike except for using different variables of quantification are equivalent and, accordingly, interchangeable.

Equivalences for Negation

QN: $\sim(\forall x)(Fx) \leftrightarrow (\exists x)(\sim Fx)$
$\sim(\exists x)(Fx) \leftrightarrow (\forall x)(\sim Fx)$

QN': $\sim(\forall x)(Fx \to Gx) \leftrightarrow (\exists x)(Fx \land \sim Gx)$
$\sim(\exists x)(Fx \land Gx) \leftrightarrow (\forall x)(Fx \to \sim Gx)$

Other Equivalences

$[(\forall x)(Fx) \land (\forall x)(Gx)] \leftrightarrow (\forall x)(Fx \land Gx)$
$[(\exists x)(Fx) \lor (\exists x)(Gx)] \leftrightarrow (\exists x)(Fx \lor Gx)$

Rules of Passage

Where p is any sentence that does *not* contain x,

(1) $[p \land (\forall x)(Fx)] \leftrightarrow (\forall x)(p \land Fx)$
(2) $[p \land (\exists x)(Fx)] \leftrightarrow (\exists x)(p \land Fx)$
(3) $[p \lor (\forall x)(Fx)] \leftrightarrow (\forall x)(p \lor Fx)$
(4) $[p \lor (\exists x)(Fx)] \leftrightarrow (\exists x)(p \lor Fx)$
(5) $[p \to (\forall x)(Fx)] \leftrightarrow (\forall x)(p \to Fx)$
(6) $[p \to (\exists x)(Fx)] \leftrightarrow (\exists x)(p \to Fx)$
(7) $[(\exists x)(Fx) \to p] \leftrightarrow (\forall x)(Fx \to p)$
(8) $[(\forall x)(Fx) \to p] \leftrightarrow (\exists x)(Fx \to p)$

Problems for Chapter 4

Parts I and II of these problems can be done after pages 145 to 167 of chapter 4 have been studied. Parts III, IV, and V are best done at the end, after the sections on negation and the rules of passage have been read; the last page of this chapter, on strategy in deduction, should, I hope, prove useful.

I. Show by deduction that the following arguments are valid:

1. All wasps are unfriendly.
   All puppies are friendly.
   Therefore, no puppies are wasps.

2. All freshman use computers.
   Only the well-educated use computers.
   Therefore, all freshman are well educated.

3. No kiwis can fly.
   Some birds are kiwis.
   Therefore, some birds cannot fly.

4. A prudent man shuns hyenas.
   No banker is imprudent.
   Therefore, no banker fails to shun hyenas.

5. No misers are unselfish.
   None but misers save eggshells.
   Therefore, everyone who saves eggshells is selfish.

6. All freshman are required to take English composition.
   Mary Jones is a freshman and an accomplished novelist.
   Therefore, some freshmen are required to take English composition.

7. No birds, except peacocks, are proud of their tails.
   Some birds, that are proud of their tails, cannot sing.
   Therefore, some peacocks cannot sing.

8. All my uncles are either very generous or good company.
   Everyone who is generous is good company.
   Therefore, all my uncles are good company.

II. In the following invalid arguments, correct one sentence (either premise or the conclusion) or else add one short missing premise, to yield a valid argument; then do a deduction. Explain why your change was needed.

1. Everyone who has a gun is a social menace.
   Some teenagers do not constitute a social menace.
   Therefore, some gunowners are not teenagers.

2. No frogs are poetical.
   Some ducks are not poetical.
   Therefore, some ducks are not frogs.

3. No frogs are poetical.
   All swans are poetical.
   Therefore, some swans are not frogs.

4. Everyone who is wise walks on his feet.
   Everyone who is unwise walks on his hands.
   Therefore, nobody walks on both.

5. Only philosophy majors may take this seminar.
   Mary and George are both philosophy majors.
   Therefore, Mary and George may take this seminar.

6. Many who use spelling checkers never learn to spell.
   Many privileged students use spelling checkers.
   Therefore, some students do not learn to spell.

III. Establish the following theorems by deduction:

1. $(\exists x)(Fx \wedge Gx) \rightarrow [(\exists x)(Fx) \wedge (\exists x)(Gx)]$
2. $[(\forall x)(Fx) \vee (\forall x)(Gx)] \rightarrow (\forall x)(Fx \vee Gx)$

Establish the following equivalences by deduction:

3. $(\exists x)(Fx \vee Gx) \leftrightarrow [(\exists x)(Fx) \vee (\exists x)(Gx)]$
4. $[(\forall x)(Fx \wedge Gx) \leftrightarrow [(\forall x)(Fx) \wedge (\forall x)(Gx)]$
5. $\sim(\exists x)(Fx \wedge Gx) \leftrightarrow (\forall x)(Fx \rightarrow \sim Gx)$
6. $\sim(\forall x)(Fx \rightarrow Gx) \leftrightarrow [(\exists x)(Fx \wedge \sim Gx)]$

IV. Show by deduction that the following arguments are valid:
1. Freshmen may be on the basketball team if and only if they have good grades.
   Some members of the basketball team do not have good grades.
   Therefore, some members of the basketball team are not freshmen.
2. All basketball players are either very tall or very fast on their feet.
   All very tall basketball players are also very fast on their feet.
   Therefore, all basketball players are fast on their feet.
3. All tall basketball players are fast on their feet.
   Some basketball players are tall or else are fast on their feet.
   Therefore, some basketball players are fast on their feet.
4. All fishermen are generous.
   No judges are generous.
   Some judges are fishermen.
   Therefore, some housewives are generous.
5. If the company goes bankrupt, everyone will be fired.
   Therefore, if the company goes bankrupt, Joe will be fired.
6. Everyone will vote for Henry or Alice will be disappointed.
   Therefore, Joe will vote for Henry or Alice will be disappointed.
7. Everyone who has a gun is a social menace.
   Therefore, if someone has a gun, someone is a social menace.
8. If someone has a gun everyone is in danger.
   Therefore, everyone who has a gun is in danger.
9. If everyone who witnessed the crime either told the truth or did not tell the truth, the criminal will be apprehended.
   Therefore, the criminal will be apprehended.

10. There is someone such that everyone will be pleased if he comes to the meeting.
There is someone such that everyone will be surprised if she is pleased.
Therefore, there is someone such that everyone will be surprised if he/she comes to the meeting.

(Disregard the "he/she" business – it is just for fun.)

V. Establish each of the following three rules of passage by deduction:

1. $[p \wedge (\forall x)(Fx)] \leftrightarrow (\forall x)(p \wedge Fx)$
2. $[p \wedge (\exists x)(Fx)] \leftrightarrow (\exists x)(p \wedge Fx)$
3. $[p \rightarrow (\forall x)(Fx)] \leftrightarrow (\forall x)(p \rightarrow Fx)$

4. *Continue* your deduction of problem 3 in order to establish:

$$[p \vee (\forall x)(Fx)] \leftrightarrow (\forall x)(p \vee Fx)$$

Assuming that your last step was, say,

21 $[p \rightarrow (\forall x)(Fx)] \leftrightarrow (\forall x)(p \rightarrow Fx)$

without any stars, your next steps might be:

22 $[\sim p \rightarrow (\forall x)(Fx)] \leftrightarrow (\forall x)(\sim p \rightarrow Fx)$
23 $[p \vee (\forall x)(Fx)] \leftrightarrow (\forall x)(p \vee Fx)$
QED

Explain and justify those two steps – or correct them.

5. Finally, do an independent deduction, by dilemma, of that same equivalence.

Chapter 5
Truth Trees for Predicate Logic

We saw in chapter 2 that truth trees could provide a yes-or-no answer to the question: Is this truth-functional argument valid? We now extend the truth-tree method to yield a similar test for predicate-logical arguments. The method will again be indirect, and the trees will look very much like the truth trees of chapter 2. The items that appear on these trees will be singular statements or negations of singular statements; where quantifiers are involved, quantified statements will be replaced by their instances.

For this purpose two new rules will be needed, to define clearly the different ways in which this replacement is to be accomplished. Existential statements must be replaced in one way, universals in another.

Truth trees have been designed as diagrams to picture the diverse ways in which a statement or conjunction of statements can turn out to be true. What an existential tells us is that there is at least one item such that the open sentence of that existential is true of it; that is, at least one instance of that existential is true. Accordingly, the content of an existential statement will be fully represented on a truth tree if we replace it by *one* of its instances, provided that the pseudoname used in the replacement is *new*.

The reader will recognize this proviso as the restriction that governs the rule of inference EI. The point is that names – whether names of people or places or objects in the world, or names of people or places or objects in novels or short stories, or pseudonames in limited logical contexts – must be so selected and so used as not to create confusion. A thing may have many names, but a name may not apply to more than one thing – at least within a given context of use. There are, after all, thousands of "Mary"s in the world, but if more than one of them turn up in a family or an office or a classroom we devise some mechanism or other to make clear which one of them we are talking about. In the limited contexts of quantificational logic we must avoid such opportunities for confusion altogether.

An existential statement to be entered on a truth tree will, accordingly, be replaced by one instance of it; the variable of quantification will be replaced by a pseudoname that is *new*, a letter which does not appear in the premises or conclusion of the argument being tested and which has not been introduced in entering any other existential.

A universal statement, on the other hand, will be dealt with differently, because it is quite different in meaning. What a universal tells us is that its open sentence is true of *everything*. So if we are talking about Mary it is true of Mary, and if we are talking about the Hudson River it is true of the Hudson River. And if we are talking about both Mary and the Hudson River it is true of both of them. When we enter a replacement for a universal on a branch of a truth tree we try to avoid naming (introducing new names), and we enter instead a succession of instances, replacing the variable of quantification by each name or pseudoname that already appears on that branch of the tree. All the relevant instances are understood to be conjoined, for that is what the universal means. A universal is fully represented on an open branch when it has been replaced by the complete list of its relevant instances.

Occasionally we have no choice but to begin with a universal, in a context in which there are no names or pseudonames to which to apply it. In such a case we do introduce a new pseudoname, but we do so only once.

It should be remembered that, if a name or pseudoname occurs twice (or more often) in a given context, we are referring twice (or more often) to the same thing, but, if two different names or pseudonames occur in the same context, their difference is no guarantee that their references are different; 'a' and 'b' may well refer to the very same thing.

Here are the two new truth-tree rules:

ET: In place of an existentially quantified sentence or component of a sentence, to be entered on an open branch of a truth tree, enter an instance of it; in each case enter a *new* name.

UT: In place of a universally quantified sentence or component of a sentence, to be entered on an open branch of a truth tree, enter every instance of it that uses a name or pseudoname already appearing on the branch in question.

Where there is choice, use ET before UT. Where there is choice, enter the "real names" first.

The reader will notice the relation of these two rules to EI and UI. He will notice also a significant difference: each of the truth-tree rules includes the phrase "or component of a sentence." In deduction EI and UI can be used only if the quantifier to be dropped covers the whole sentence that is being instantiated. ET and UT work on components as well – but the trees must be carefully marked to make clear how wide was the scope of the eliminated quantifier.

So, '$(\exists x)(Fx \land Gx)$' might yield

$$\begin{matrix} Fa \\ Ga \end{matrix} \Big] \quad \text{PREM 1, ET } (a)$$

and $(\exists y)(Fy \lor Gy)$ might yield

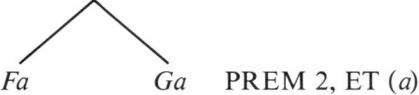

$$\quad Fa \qquad Ga \quad \text{PREM 2, ET } (a)$$

In contrast, where two existential quantifiers, relatively narrow in scope, occur, ET must be used twice, and so we have two new pseudonames. '$(\exists x)(Fx) \land (\exists x)(Gx)$' might yield

$$\begin{matrix} Fa & \big| & \text{ET } (a) \\ \\ Gb & \big| & \text{ET } (b) \end{matrix} \Bigg] \quad \text{PREM 3}$$

and '$(\exists x)(Fx) \lor (\exists x)(Gx)$' might yield

```
          ╱╲
ET (a)  Fa    Gb  ET (b)   PREM 4
```

As we know, $(\exists x)(Fx \land \sim Fx)$ is inconsistent:

$$\left.\begin{array}{r} Fa \\ \sim Fa \end{array}\right| \; \Big] \; \text{ET } (a)$$
X

whereas $(\exists x)(Fx) \land (\exists x)(\sim Fx)$ is not:

$$\left.\begin{array}{r|l} Fa & \text{ET } (a) \\ \sim Fb & \text{ET } (b) \end{array}\right] \; \text{PREM 5}$$
↑

These two new rules, ET and UT, make no mention of negation. So the denials of quantified sentences or clauses must be handled as the denials of compound sentences and their components were handled with respect to truth-functional truth trees: by internalizing negation. To diagram the denial of a quantified conclusion we first internalize negation and then use either either ET or UT, as appropriate. For internalizing negation we rely on the two pairs of equivalences for negation introduced in chapter 4:

QN $\sim(\forall x)(Fx) \leftrightarrow (\exists x)(\sim Fx)$
    $\sim(\exists x)(Fx) \leftrightarrow (\forall x)(\sim Fx)$

QN' $\sim(\forall x)(Fx \to Gx) \leftrightarrow (\exists x)(Fx \land \sim Gx)$
     $\sim(\exists x)(Fx \land Gx) \leftrightarrow (\forall x)(Fx \to \sim Gx)$

It is worth while to pay special attention again, as we did in chapter 2, to the problem of "implicit negation." How, for example, do we diagram a conditional with quantified antecedent? We do just what we did in chapter 2 when we wanted to diagram a conditional with compound antecedent. We remind ourselves of the basic diagram for IF:

Truth Trees for Predicate Logic   191

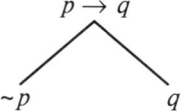

and we internalize negation in the denial of the antecedent before we build the truth tree. For example,

> If someone is hungry or thirsty, Joe will go to the store to get supplies.
> Clara is hungry.
> So Joe will go to the store.
>
> PREM 1: $(\exists x)(Hx \lor Tx) \to Sj$
> PREM 2: Hc
> ~CON: $\sim Sj$
> PREM1: $\sim(\exists x)(Hx \lor Tx) \lor Sj$ by IF
> $(\forall x) \sim (Hx \lor Tx) \lor Sj$ by QN
> $(\forall x)(\sim Hx \land \sim Tx) \lor Sj$ by NOR

Hc | PREM 2

$\sim Sj$ | ~CON

UT $\begin{bmatrix} \sim Hc \\ \sim Tc \\ \sim Hj \\ \sim Tj \end{bmatrix}$ $\begin{matrix} Sj \\ \mathbf{X} \end{matrix}$ PREM 1

X   VALID

This tree is "complete"; the denial of the antecedent has been applied to both Clara and Joe, and the branch has dealt with both hunger and thirst. All this would have been needed if the argument had been invalid, and it is important to know it is available, but in this case the branch closed after the first entry, so this little tree would have been sufficient:

192   Predicate Logic

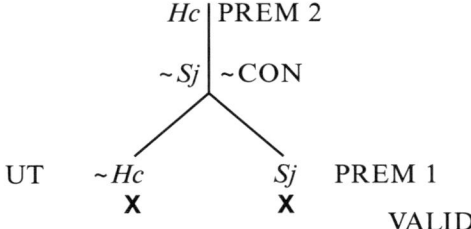

Another pair of examples is worth considering, in part so as to make quite clear the difference in meaning between the two sentences involved and to emphasize the importance of getting the scopes of the quantifiers right.

> If everyone has a gun, everyone is in danger.
> Therefore, everyone who has a gun is in danger.

Intuitively, reading this argument in English, one tends to view the premise of the argument as true, but its conclusion as false; so one might guess that the argument is invalid. This guess is of course correct, but getting the truth tree right depends on correct handling of the scopes of the quantifiers and correct use of the equivalences for negation.

```
   PREM:  (∀x)(Gx) → (∀x)(Dx)
          ~(∀x)(Gx) ∨ (∀x)(Dx)            by IF
          (∃x)(~Gx) ∨ (∀x)(Dx)            by QN
    CON:  (∀x)(Gx → Dx)
   ~CON:  (∃x)(Gx ∧ ~Dx)                  by QN'
```

                       Ga ⎤
                      ~Da ⎦   ~CON, ET (a)
                      ╱ ╲
        ╱   ╱   ╲
ET (b) ~Gb   Da   UT   PREM
         ↑   X   INVALID

The converse argument, on the other hand, is valid:

Everyone who has a gun is in danger.
Therefore, if everyone has a gun, everyone is in danger.

PREM: $(\forall x)(Gx \to Dx)$
CON: $(\forall x)(Gx) \to (\forall x)(Dx)$
~CON: $(\forall x)(Gx) \land \sim(\forall x)(Dx)$ by NIF
        $(\forall x)(Gx) \land (\exists x)(\sim Dx)$ by QN

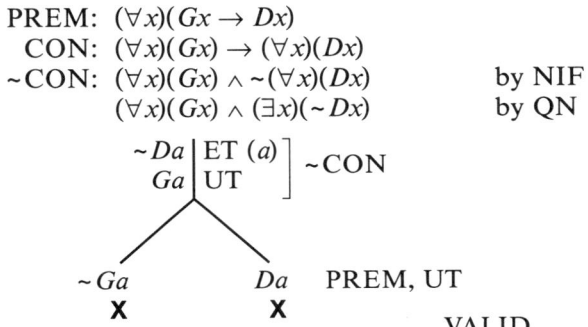

Let us turn next to the two syllogisms we proved in introducing EI and UG.

Some burglars are caught by the police.
All burglars are dishonest.
Therefore, some who are dishonest are caught by the police.

PREM 1: $(\exists x)(Bx \land Cx)$
PREM 2: $(\forall x)(Bx \to Dx)$
CON: $(\exists x)(Dx \land Cx)$
~CON: $(\forall x)(Dx \to \sim Cx)$ by QN'

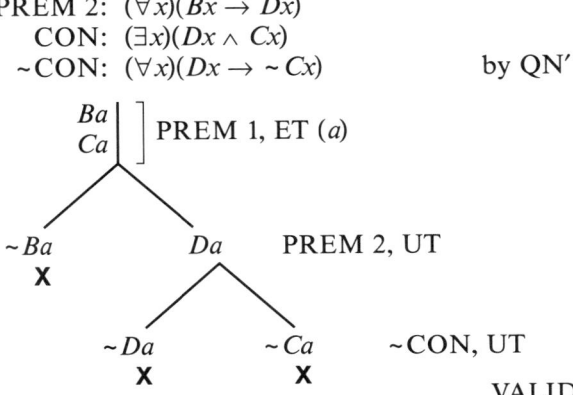

Next: All winged things can fly.
       All birds have wings.
       Therefore, all birds can fly.

194   Predicate Logic

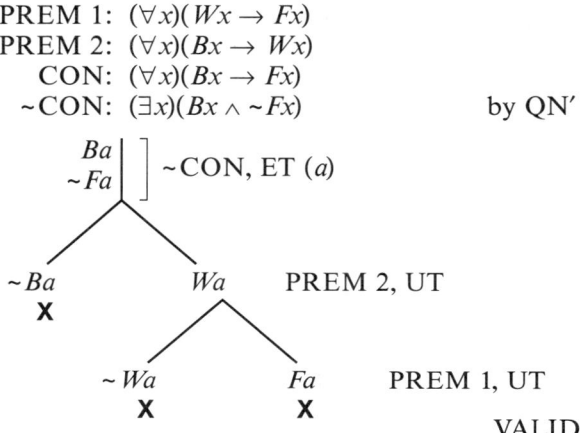

Note that, whereas the order of the two premises in this case does not matter, the denial of the conclusion should be entered first. This is a matter of convenience, not of principle. The following diagram shows how a tree built with the existential used after the universals would look:

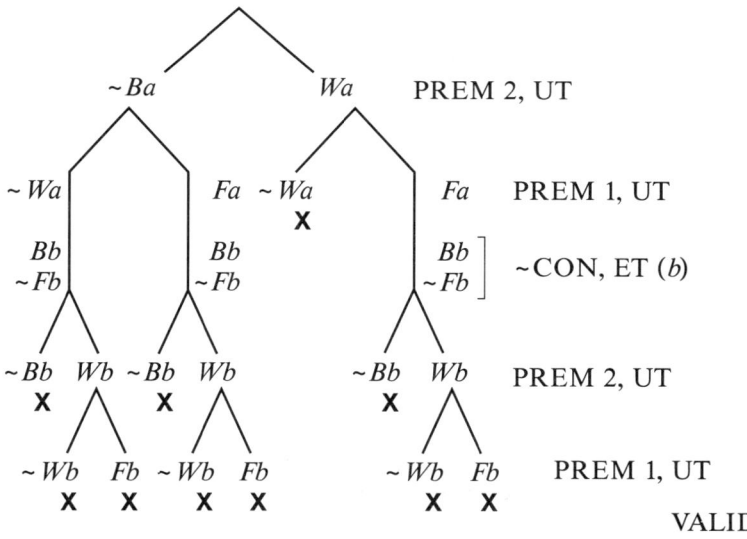

This works, provided one is careful to reuse the universal premises, for they apply to b as well as a – but it is clumsy.

We turn next to the invalid arguments we used in chapter 4 to illustrate misuse of EI and UG.

A. Someone is in the closet and wants to get out.
   Therefore, Joe Jenkins is in the closet.

   PREM: $(\exists x)(Cx \wedge Wx)$
   ~CON: ~Cj

   $\quad\quad \left|\begin{array}{l} \sim Cj \\ \left.\begin{array}{l} Ca \\ Wa \end{array}\right] \\ \uparrow \end{array}\right.$ $\quad$ ~CON

   $\quad\quad\quad\quad\quad\quad$ PREM, ET (a)

   $\quad\quad\quad\quad\quad\quad$ INVALID

Three points should be noted. First, Ca (Albert, say, is in the closet) and ~Cj (Joe Jenkins is not in the closet) are not contradictory; so the little tree is open. Secondly, the open tree provides a counterexample and so makes clear why the original argument was invalid: somebody, say Albert, is in the closet and wants to get out, but Joe Jenkins is not in the closet – given the premise, this is quite possible. Thirdly, order of work is again important here; a second recommendation is being used:

> Where possible, enter the "real names", or names that occur in the premises or the conclusion of the argument being tested, before use of ET or UT.

If ET or UT were used first, one might be tempted to introduce a pseudoname inappropriately: that is, matching a "real name" if we are using ET (in such cases as this), or failing to match a real name if we were using UT.

B. If Mary eats strawberries, Mary gets a rash.
   Therefore, everyone who eats strawberries gets a rash.

196   Predicate Logic

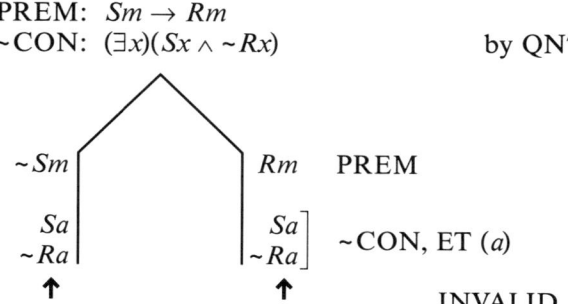

As in chapter 2, equivalences for negation are used for internalizing negation before items are entered on the tree.

C. Some animals are cats and some animals are dogs. Therefore, some cats are dogs.

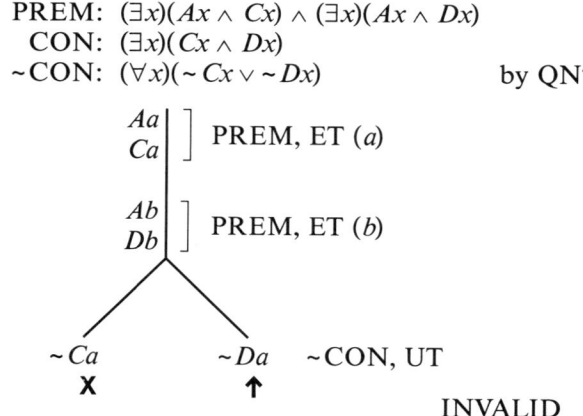

The tree shown above is *not* correct. The arrow on the right-hand branch and the notation "INVALID" are both premature. They are premature because the denial of the conclusion has been used only once, as applied to a; but b also appears on the open branch. The complete tree looks like this:

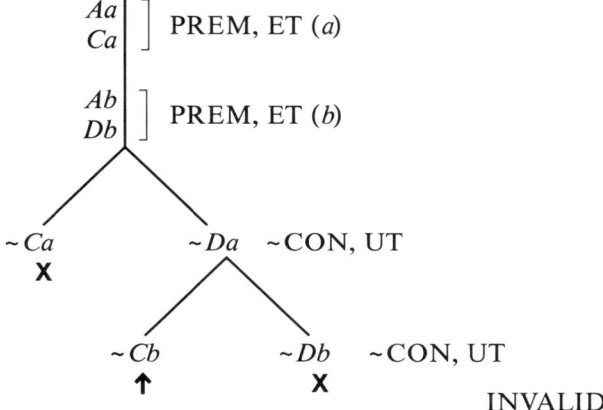

The open branch now provides the counterexample that explains the invalidity of the argument: It is possible, given the premise, that Algernon, who is an animal and a cat, is not a dog, whereas Bertrand, who is an animal and a dog, is not a cat; this arrangement would satisfy the premise, but is compatible with the falsity of the conclusion. The incomplete tree provides only half that information. Also, there are cases in which a second use of UT is needed in order to close the tree for an argument that will turn out to be valid; in such cases the use of an incomplete tree could be misleading, for it could lead one to take a valid argument to be invalid.

Note also the structure of the diagram for $(\forall x)(Fx \vee Gx)$ as applied to two pseudonames. $(\forall x)(Fx \vee Gx)$ tells us that $(Fa \vee Ga)$ and $(Fb \vee Gb)$, or

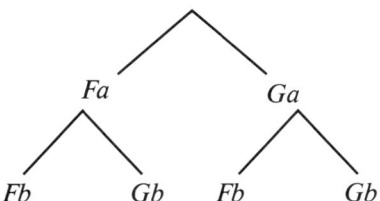

This is quite different from

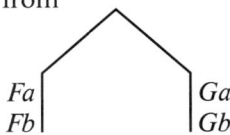

which is the diagram for $(Fa \wedge Fb) \vee (Ga \wedge Gb)$ or for $(\forall x)(Fx) \vee (\forall x)(Gx)$. Unlike argument C, this argument:

   C'. Some animals are dogs and some animals are cats. Therefore, there are dogs and there are cats.

is valid (though uninteresting).

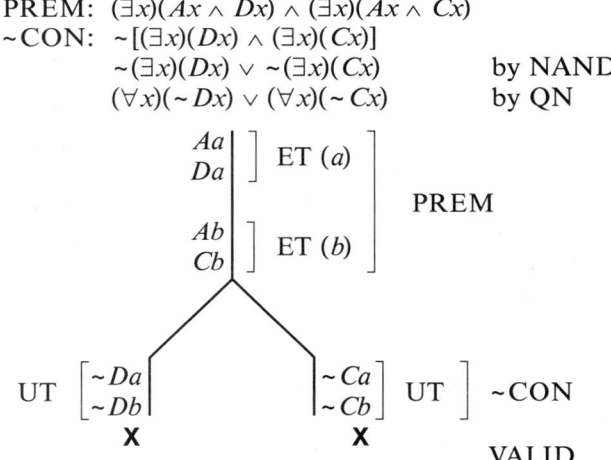

D. Something is interesting.
   Therefore, everything is interesting.

```
PREM:   (∃x)(Ix)
 CON:   (∀x)(Ix)
~CON:   (∃x)(~Ix)                by QN
```

$$\begin{array}{c|l} Ia & \text{PREM, ET }(a) \\ \\ \sim Ib & \sim \text{CON, ET }(b) \\ \uparrow & \quad \text{INVALID} \end{array}$$

The contrasting valid argument D′ is an example of the use of UT to introduce a pseudoname when there is no alternative. D′ was

> Everything is interesting.
> Therefore, something is interesting.
>
> PREM: $(\forall x)(Ix)$
> ~CON: $~(\exists x)(Ix)$
> $(\forall x)(\sim Ix)$ by QN
>
> Ia | PREM, UT
>
> $\sim Ia$ | ~CON, UT
> **X**
>   VALID

This argument, as was already clear from the deduction in chapter 4, reflects the metaphysical assumption noted in our initial discussion of UI, namely that there is something, that we are working in a nonempty universe.

A "related" argument, however, is invalid.

> D″ All magicians are interesting.
> Therefore, some magicians are interesting.
>
> PREM: $(\forall x)(Mx \to Ix)$
> CON: $(\exists x)(Mx \wedge Ix)$
> ~CON: $(\forall x)(Mx \to \sim Ix)$ by QN′

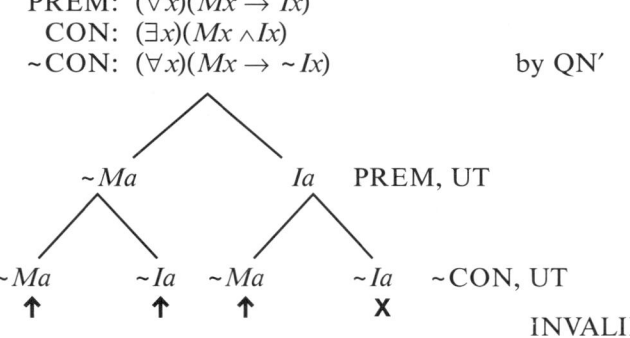

INVALID

We do *not* make the "metaphysical" assumption that there are magicians.

E. Everything is interesting or uninteresting.
Therefore, everything is interesting or everything is uninteresting.

PREM: $(\forall x)(Ix \vee \sim Ix)$
CON: $(\forall x)(Ix) \vee (\forall x)(\sim Ix)$
~CON: $\sim[(\forall x)(Ix) \vee (\forall x)(\sim Ix)]$
       $\sim(\forall x)(Ix) \wedge \sim(\forall x)(\sim Ix)$ by NOR
       $(\exists x)(\sim Ix) \wedge (\exists x)(Ix)$ by QN, DN

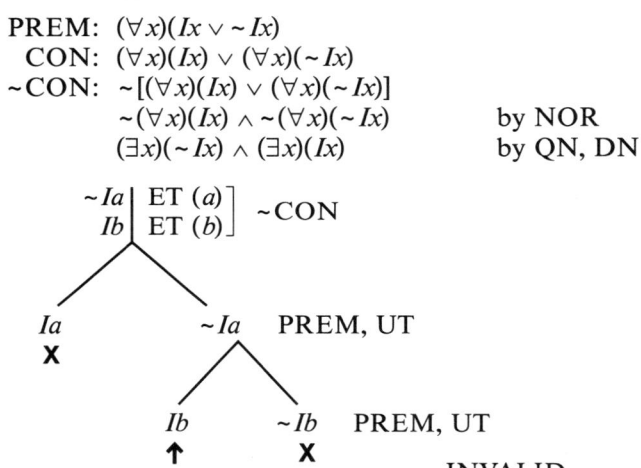

INVALID

Note that the flags I have been using with ET are not really needed, but they are useful reminders; each use of ET introduces a new pseudoname. If any names or pseudonames are already on the tree, use of UT does not introduce any new pseudonames; UT applies to what is there.

One must remind oneself to use UT often enough, and to mark one'e trees clearly, so that a reader can see easily where each item comes from.

Some Problems in Predicate Logic

Let us look now at a few slightly more complicated examples, noting, as we work, how one would go about arriving at a preliminary estimate of their validity.

F. All freshmen must take either mathematics or science.
Not all freshmen take mathematics.
Therefore, some who take science do not take mathematics.

$(\forall x)[(Fx \rightarrow (Mx \vee Sx)]$
$\sim(\forall x)(Fx \rightarrow Mx)$
∴ $(\exists x)(Sx \wedge \sim Mx)$

This *seems* valid; for the second premise tells us (if we remember QN') that some freshmen do not take mathematics, and the first premise tells us that those freshmen, since they don't take mathematics, must take science. We have found what the conclusion calls for: some who take science but not mathematics. The above is a sketch for a deduction. The truth tree looks like this:

PREM 1: $(\forall x)[Fx \rightarrow (Mx \vee Sx)]$
PREM 2: $\sim(\forall x)(Fx \rightarrow Mx)$
    $(\exists x)(Fx \wedge \sim Mx)$ by QN'
CON: $(\exists x)(Sx \wedge \sim Mx)$
~CON: $(\forall x)(Sx \rightarrow Mx)$ by QN'

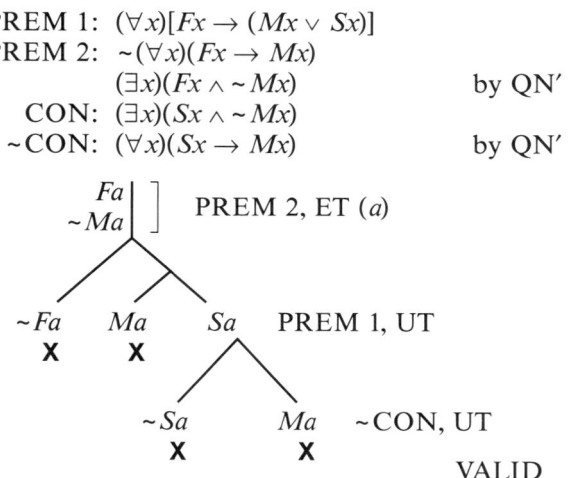

Predicate Logic

G. Some Greeks are judges.
Some Greeks are kings.
All judges are fair-minded.
All kings are haughty.
Therefore, some fair-minded Greeks are haughty.

PREM 1: $(\exists x)(Gx \land Jx)$
PREM 2: $(\exists x)(Gx \land Kx)$
PREM 3: $(\forall x)(Jx \to Fx)$
PREM 4: $(\forall x)(Kx \to Hx)$
CON: $(\exists x)(Fx \land Gx \land Hx)$
~CON: $(\forall x)(\sim Fx \lor \sim Gx \lor \sim Hx)$ by QN, NAND

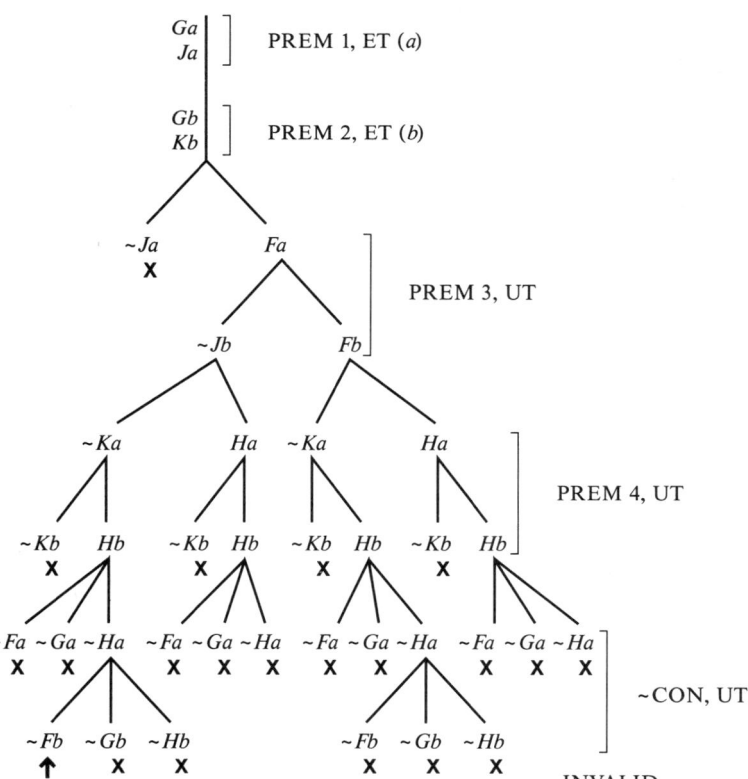

Is this argument valid? Before we look at its truth tree (opposite, on page 202), let us consider this question. The first premise tells us that some Greeks are judges and the third that those judges are fair-minded; so some Greeks are fair-minded. The second premise tells us that some Greeks are kings, and the fourth that those kings are haughty; so some Greeks are haughty. But do the premises give us reason to believe that these two groups of Greeks have members in common? They do not. We have tried to work out a sketch for a deduction, but that attempt has failed. We estimate that the argument is invalid and that the tree will have an open branch. It does have.

The open branch, which reads: *Ga, Ja, Gb, Kb, Fa,* ~*Jb,* ~*Ka, Hb,* ~*Ha,* ~*Fb,* shows that, given the premises, there may be a Greek judge, Aristides (*a*), who is fair-minded, but who is neither a king nor haughty, and a Greek king, Brasidas (*b*), who is haughty, but who is neither a judge nor fair-minded, and that this is possible even if the premises are true and the conclusion is false, that is, even if no fair-minded Greeks are haughty. The argument is invalid. (To establish invalidity, it was necessary to finish only the left-hand section of this tree.)

> H. All freshmen have computers.
> If no one who has a computer can spell then everyone who has a computer uses a spelling checker.
> Therefore, if any freshmen do not use spelling checkers, some computer owners can spell.

This argument presents three problems in formalization, the first (and easiest to manage) having to do with setting up abbreviations that avoid confusion, the second having to do with the scope of quantifiers, and the third dealing with the word 'any'.

> Let $Fx = x$ is a freshman.
> Let $Ox = x$ is a computer owner.
> Let $Sx = x$ can spell.
> Let $Ux = x$ uses a spelling checker.

Next, it is essential to notice that both the second premise and the conclusion are conditional statements, with quantified antecedents and quantified consequents. So any impulse to formalize either of them with one quantifier covering the whole statement is to be resisted. The antecedent of the second premise should read either '$\sim(\exists x)(Ox \wedge Sx)$' or '$(\forall x)(Ox \to \sim Sx)$' – in either case a closed formula. The consequent of that premise would then be '$(\forall x)(Ox \to Ux)$'. Some such reading for the second premise as

$$(\forall x)[(Ox \to \sim Sx) \to (Ox \to Ux)]$$

would be quite misleading. That misformulation is not equivalent to the correct

$$(\forall x)(Ox \to \sim Sx) \to (\forall x)(Ox \to Ux)$$

Other possibilities are:

$$\sim(\exists x)(Ox \wedge Sx) \to (\forall x)(Ox \to Ux)$$
$$\sim(\exists x)(Ox \wedge Sx) \to \sim(\exists x)(Ox \wedge \sim Ux)$$
$$(\forall x)(Ox \to \sim Sx) \to (\forall y)(Oy \to Uy)$$

The change of variable in the last reading is helpful in that it emphasizes the narrowness of the scopes and the consequent absence of cross reference. Stretching the scopes of both quantifiers on this basis, as follows:

$$(\exists x)(\forall y)[(Ox \to \sim Sx) \to (Oy \to Uy)]$$

or

$$(\forall y)(\exists x)[(Ox \to \sim Sx) \to (Oy \to Uy)]$$

would be legitimate, but not particularly helpful. Notice the contrast between these formulas and the incorrect

$$(\forall x)[(Ox \to \sim Sx) \to (Ox \to Ux)]$$

suggested (and rejected) above.

It is worth noting that, whereas wide scope is sometimes convenient in deduction, where EI or UI is to be used, narrow scope is usually convenient for diagramming by truth tree.

We next turn to the conclusion, which can be read:

> If some freshmen do not use spelling checkers, some computer owners can spell.

where the word 'some' has replaced the 'any' of the original. The likeness of meaning will be intuitively evident, I think, to most

English speakers, but it is worth while to give some thought to how it works. The word 'any' (along with such compounds as 'anyone'. 'anything', 'anywhere') has the force of universality, but that force is conveyed by different replacements in various contexts. In the affirmative, 'any' can be replaced by 'every' or 'all':

> I enjoy any outdoor activity.
> If anything is an outdoor activity I enjoy it.
> I enjoy all outdoor activity.
> I enjoy every outdoor activity.
>
> $(\forall x)(Ox \rightarrow Ex)$

But, in the negative, 'any' must be replaced not by 'all' or 'every', but by (if anything) the (sometimes unidiomatic) word 'some':

> I do not enjoy any outdoor activity.
> It is not the case that I enjoy some outdoor activity.
> It is not the case that there is any outdoor activity that I enjoy.
> It is not the case that there is some outdoor activity that I enjoy.
>
> $\sim(\exists x)(Ox \land Ex)$

And the result does convey universality, for it is equivalent to

> $(\forall x)(Ox \rightarrow \sim Ex)$

Again, as we saw in connection with (7) of the rules of passage, the following are equivalent:

> If anyone pushes the button, the elevator comes up.
> If someone pushes the button, the elevator comes up.
>
> $(\forall x)(Px \rightarrow E)$
> $(\exists x)(Px) \rightarrow E$ by (7) of the rules of passage

To summarize, where 'any' occurs either in the negative or as conveying the "quantity" of the antecedent of a conditional, it is symbolized by the existential quantifier. Otherwise it is symbolized by a universal quantifier.

So we arrive at the following formalization of argument H:

Predicate Logic

PREM 1 $(\forall x)(Fx \to Ox)$
PREM 2 $\sim(\exists x)(Ox \wedge Sx) \to (\forall x)(Ox \to Ux)$
          $(\exists x)(Ox \wedge Sx) \vee (\forall x)(Ox \to Ux)$ by IF
    CON $(\exists x)(Fx \wedge \sim Ux) \to (\exists x)(Ox \wedge Sx)$
  ~CON $(\exists x)(Fx \wedge \sim Ux) \wedge \sim(\exists x)(Ox \wedge Sx)$ by NIF
          $(\exists x)(Fx \wedge \sim Ux) \wedge (\forall x)(Ox \to \sim Sx)$ by QN′

But, before building the truth tree, let us see whether this argument *seems* valid. Let us assume that some freshmen do not use spelling checkers (we are sketching a possible conditional proof, and this is the antecedent of the desired conclusion). Can we show that, in that case, some computer owners can spell? Since the first premise tells us that all freshmen have computers and we have assumed that some freshman do not use spelling checkers, it follows that some who have computers (freshmen) do not use spelling checkers. But this tells us that the consequent of the second premise (everyone who has a computer uses a spelling checker) is false, and so also (by modus tollens) must be its antecedent. That is, it is not the case that no one who has a computer can spell. The antecedent of the conclusion has led to a statement equivalent to the consequent of the conclusion, and so our judgment is that the argument is valid.

But this result must be checked both by truth tree and, if it is confirmed, by formal deduction.

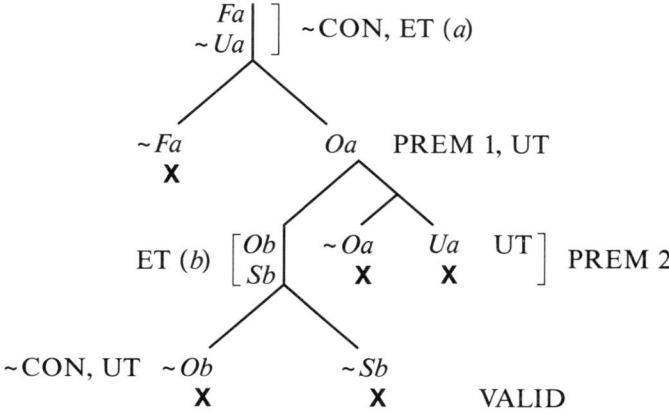

Truth Trees for Predicate Logic 207

It should be noted that, whereas only the pseudoname '*a*' appears on the right-hand side of this tree, where UT is used twice, in contrast, on the left-hand side, where the replacement for '$(\forall x)(Ox \rightarrow \sim Sx)$' is entered, that universal applies to *both a* and *b*. Instead of

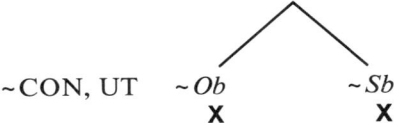

at the bottom of the truth tree, I could have shown:

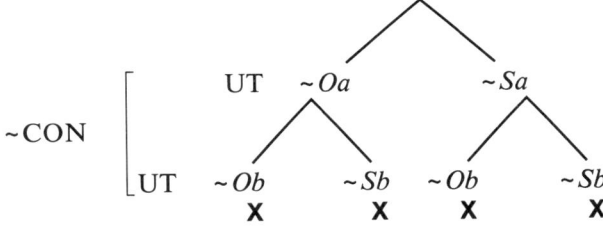

This would have yielded the more complicated diagram:

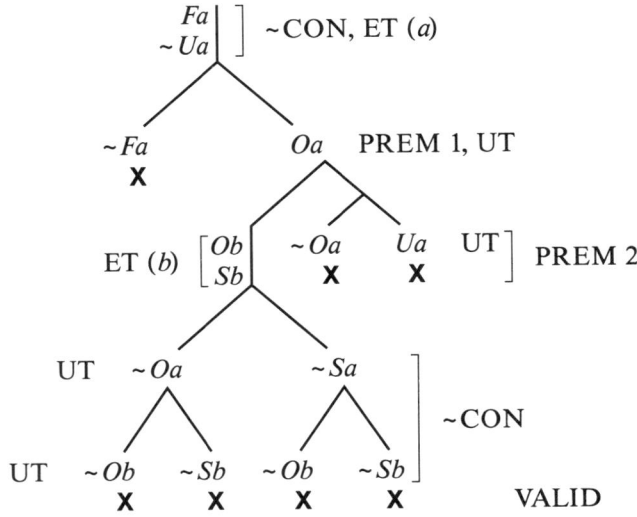

Explicit use of both "$Oa \to \sim Sa$" and "$Ob \to \sim Sb$" would have been required if we were establishing invalidity, but here, by choosing to use 'b' first, before 'a', we close the branch promptly and shorten the tree; mention of a on the left-hand side turns out to be irrelevant.

The formal deduction should be done by the reader, following the informal sketch above or, if he prefers, by *reductio ad absurdum* or some other pattern of proof.

Nested Quantifiers

We deal next with a pair of statements, to be tested for validity, both of which involve *nested quantifiers* and so require use of two variables, for clarity. The reader will remember from chapter 2 that, to test a *statement* for validity, we test its negation for inconsistency. We take notice first, however, of the special problem of formalization which is involved in nesting, that is, the occurrence of quantifiers within the scopes of other quantifiers.

J. Everyone is such that, if he is foolish, everyone is.

We are to decide whether this statement is valid. Here is a first attempt at formalization:

$(\forall x)[Fx \to (\forall x)(Fx)]$

This formula is hard to read, since the second 'Fx' is covered by two x quantifiers, and we cannot see immediately which of them applies. We recall that, by change of variable, CV, of chapter 4, $(\forall x)(Fx)$ can be interchanged with '$(\forall y)(Fy)$'. So we use instead:

J $(\forall x)[Fx \to (\forall y)(Fy)]$

'$(\forall y)[Fy \to (\forall x)(Fx)]$', of course, would do as well.

Is this statement valid? Think of some foolish person you know, say Joe; is it true of Joe that if he is foolish, everyone is? That is, since Joe *is* foolish, is everyone foolish? Probably not. The statement *seems* invalid.

Note also that we can apply rule (7) of the rules of passage:

$[(\exists x)(Fx) \to p] \leftrightarrow (\forall x)(Fx \to p)$

Substituting '$(\forall y)(Fy)$' for 'p', we obtain:

$[(\exists x)(Fx) \to (\forall y)(Fy)] \leftrightarrow (\forall x)[Fx \to (\forall y)(Fy)]$

and, accordingly, J can be rewritten as

J' $(\exists x)(Fx) \to (\forall y)(Fy)$

If someone is foolish, everyone is foolish. And that is surely invalid. We test this intuition by truth tree:

$\sim(\forall x)[Fx \to (\forall y)(Fy)]$	
$(\exists x) \sim[Fx \to (\forall y)(Fy)]$	by QN
$(\exists x)[Fx \wedge \sim(\forall y)(Fy)]$	by NIF
$(\exists x)[Fx \wedge (\exists y)(\sim Fy)]$	by QN
$(\exists x)(Fx) \wedge (\exists y)(\sim Fy)$	by (2) of the rules of passage

We could, alternatively, have used the rules of passage before negating and started with the denial of J':

$\sim[(\exists x)(Fx) \to (\forall y)(Fy)]$	
$(\exists x)(Fx) \wedge \sim(\forall y)(Fy)$	by NIF
$(\exists x)(Fx) \wedge (\exists y)(\sim Fy)$	by QN

The conjunction to be diagrammed is the same. We obtain

Fa	ET (a)
$\sim Fb$	ET (b)
↑	INVALID

$(\forall x)[Fx \to (\forall y)(Fy)]$, as expected, is invalid. But *this* statement: $(\exists x)[Fx \to (\forall y)(Fy)]$, perhaps unexpectedly, is not.

K. There is someone such that, if he is foolish, everyone is.

$(\exists x)[Fx \to (\forall y)(Fy)]$

We start by checking this statement by truth tree. Its denial is:

$(\forall x) \sim[Fx \to (\forall y)(Fy)]$	by QN
$(\forall x)[Fx \wedge \sim(\forall y)(Fy)]$	by NIF

$(\forall x)[Fx) \land (\exists y)(\sim Fy)]$ by QN
$(\forall x) (Fx) \land (\exists y)(\sim Fy)$ by (1) of the rules of passage

    $\sim Fa$ | ET (*a*)
      Fa | UT
      X   VALID

Since its negation is inconsistent, $(\exists x)[Fx \to (\forall y)(Fy)]$ is valid. We now establish it by deduction:

 *1 Fa PREM
 *2 $(\forall y)(Fy)$ 1 UG (*a*)
  3 $Fa \to (\forall y)(Fy)$ 1–2 CP
  4 $(\exists x)[Fx \to (\forall y)(Fy)]$ 3 EG
                                                   QED

This little deduction establishes its conclusion. It does not, however, "establish" its step 3, since, without step 4, the deduction is unfinished. Step 4 eliminates the flagged pseudoname *a* from the conclusion; so step 4 cannot be omitted. Nor could it be replaced by the alternative step:

  4′ $(\forall x)[Fx \to (\forall y)(Fy)]$ 3 UG (*a*)

That step would, if it were acceptable, establish J instead of K. Step 4′ is disallowed by the restriction against double flagging – which suggests that J is indeed invalid.

Finally, we see whether the rules of passage can be used to shed light on the validity of K. Rule (8) tells us that

$$[(\forall x)(Fx) \to p] \leftrightarrow (\exists x)(Fx \to p)$$

and so K, $(\exists x)[Fx \to (\forall y)(Fy)]$, is interchangeable with

$$(\forall x)(Fx) \to (\forall y)(Fy)$$

which is obviously valid, since it is an instance of the "law of identity" ($p \to p$), with only an inconsequential change of variable. So we have a third demonstration of the validity of K: "There is someone such that if he is foolish, everyone is foolish."

Let us now try to make sense of this odd claim. Let us imagine a philosopher setting out, like Diogenes, with a lamp, looking not exactly for a wise man but for someone such that if he is foolish, everyone is foolish. Suppose the first man he meets is very foolish; since the philosopher does not know whether everyone is foolish, he does not know whether this foolish man meets his criterion. He tries again. Suppose the second person he meets is *not* foolish; in this case the philosopher has found his man; it is true of this nonfoolish person that if he is foolish, everyone is foolish (by virtue of falsity of the antecedent). So there are two possibilities: either in his wanderings the philosopher finds someone (*a*) who is not foolish, and '$Fa \to (\forall y)(Fy)$' is true by virtue of falsity of the antecedent, or else, wherever he goes, he finds only people who *are* foolish, and he eventually comes to believe that everyone is, or $(\forall y)(Fy)$, and so that '$Fb \to (\forall y)(Fy)$' is true by virtue of the truth of the consequent.

This little fable is *not* a demonstration – but we have provided three demonstrations already.

L. Let us return to an argument involving nested quantifiers which was assigned for deduction in chapter 4:

There is someone such that everyone will be pleased if he comes to the meeting.
There is someone such that everyone will be surprised if she is pleased.
Therefore, there is someone such that everyone will be surprised if he comes to the meeting.

$(\exists x)[Mx \to (\forall y)(Py)]$
$(\exists x)[Px \to (\forall y)(Sy)]$
$\therefore (\exists x)[Mx \to (\forall y)(Sy)]$

A preliminary assessment of the validity of this argument would involve, I think, sketching a deduction. There are several ways in which this could be done.

First, *as it stands*:
Suppose there is someone, say, Alfred, such that everyone is pleased if he comes to the meeting, and someone, say Beatrice, such that everyone is surprised if she is pleased. Now suppose that Alfred comes to the meeting. Everyone will be pleased, including Beatrice, and so, since Beatrice is pleased, everyone will

be surprised. Alfred (by Conditional Proof) meets the criterion of the desired conclusion, and the argument is valid. Formally,

$$* \begin{cases} 1 & (\exists x)[Mx \to (\forall y)(Py)] \\ 2 & (\exists x)[Py \to (\forall y)(Sy)] \end{cases} \quad \text{PREM}$$

To Prove: $(\exists x)[Mx \to (\forall y)(Sy)]$

*	3	$Ma \to (\forall y)(Py)$	1 EI (a)
*	4	$Pb \to (\forall y)(Sy)$	2 EI (b)
**	5	Ma	PREM
**	6	$(\forall y)(Py)$	1, 5 MP
**	7	Pb	6 UI
**	8	$(\forall y)(Sy)$	4, 7 MP
*	9	$Ma \to (\forall y)(Sy)$	5–8 CP
*	10	$(\exists x)[Mx \to (\forall y)(Sy)]$	9 EG
			QED

Second, *in pure form*:
The premises, by rule of passage (8), are equivalent to "If everyone comes to the meeting everyone will be pleased" and "If everyone is pleased everyone will be surprised." Accordingly, the chain rule will yield "If everyone comes to the meeting everyone will be pleased," which is similarly equivalent to the desired conclusion. Formally,

$$* \begin{cases} 1 & (\exists x)[Mx \to (\forall y)(Py)] \\ 2 & (\exists x)[Px \to (\forall y)(Sy)] \end{cases} \quad \text{PREM}$$

*	3	$(\forall x)(Mx) \to (\forall y)(Py)$		1 INT [by rule of passage (8)]
*	4	$(\forall x)(Px) \to (\forall y)(Sy)$	(likewise)	2 INT
*	5	$(\forall y)(Py) \to (\forall y)(Sy)$		4 INT (CV)
*	6	$(\forall x)(Mx) \to (\forall y)(Sy)$		3, 5 CHAIN
*	7	$(\exists x)[Mx \to (\forall y)(Sy)]$		6 INT
				QED

Here we have invoked rule of passage (8) three times: at steps 3 and 4, to narrow the scope of the "*x*" quantifier, as we did with respect to statement K above, and again, in the opposite direction, at step 7, to widen the scope again.

Either of the above deductions might well come to mind, as justification for one's assessment that argument L is valid; either can easily be sketched in ordinary language. Another possibility is less easy to grasp intuitively, but is formally convenient.

Third, *in prenex* form:

* $\begin{cases} 1 & (\exists x)[Mx \to (\forall y)(Py)] \\ 2 & (\exists x)[Px \to (\forall y)(Sy)] \end{cases}$ PREM
* 3 $(\exists x)(\forall y)(Mx \to Py)$ 1 INT [by rule of passage (6)]
* 4 $(\exists x)(\forall y)(Px \to Sy)$ 2 INT (likewise)
* 5 $(\forall y)(Ma \to Py)$ 3 EI (*a*)
* 6 $(\forall y)(Pb \to Sy)$ 4 EI (*b*)
* 7 $Ma \to Pb$ 5 UI
* 8 $Pb \to Sc$ 6 UI
* 9 $Ma \to Sc$ 7, 8 CHAIN
* 10 $(\forall y)(Ma \to Sy)$ 9 UG (*c*)
* 11 $(\exists x)(\forall y)(Mx \to Sy)$ 10 EG
* 12 $(\exists x)[Mx \to (\forall y)(Sy)]$ 11 INT
     QED

Here rule of passage (6) is invoked three times: at steps 3 and 4 to widen the scope of the "*y*" quantifier, and again at step 12, to narrow that scope again and restore the form of the original conclusion. As noted in chapter 4, the rules of passage – like the equivalences for negation – may be used either to narrow the scope of quantifiers or to widen that scope. When the scopes of all quantifiers are narrow, so that there is *no* nesting, we say that the formula in question is in *pure* form. When the scopes of all quantifiers are wide, all of them preceding a single open sentence (the *nexus*), we say that the sentence in question is in *prenex* form. Internalizing negation may be thought of a sort of "prenexing."

Let us turn now to the truth tree for our argument about the meeting. It will be easier to build if we do some preparation before we do the diagram.

214   Predicate Logic

PREM 1: $(\exists x)[Mx \to (\forall y)(Py)]$
1': $Ma \to (\forall y)(Py)$ by ET (a)
PREM 2: $(\exists x)[Px \to (\forall y)(Sy)]$
2': $Pb \to (\forall y)(Sy)$ by ET (b)
CON: $(\exists x)[Mx \to (\forall y)(Sy)]$
~CON: $\sim(\exists x)[Mx \to (\forall y)(Sy)]$
     $(\forall x) \sim [Mx \to (\forall y)(Sy)]$ by QN
     $(\forall x)[Mx \wedge \sim(\forall y)(Sy)]$ by NIF
     $(\forall x)[Mx \wedge (\exists y)(\sim Sy)]$ by QN
     $(\forall x)(Mx) \wedge (\exists y)(\sim Sy)$ by rule of passage (1)
~CON A: $(\forall x)(Mx)$ by SIMP
~CON B: $(\exists y)(\sim Sy)$ by SIMP
     $\sim Sc$ by ET (c)

```
                            ~Sc    ~CON B, ET (c)
                           /    \
              ET (b)  ~Pb        Sc  UT  PREM 2'
                     /    \       X
      ET (a)  ~Ma           Pb  UT  PREM 1'
                                X
   ~CON A, UT  Ma
               X                              VALID
```

Another way to handle this truth tree might be to purify at once:

PREM 1: $(\exists x)[Mx \to (\forall y)(Py)]$
     $(\forall x)(Mx) \to (\forall y)(Py)$ by rule of passage (8)
     $\sim(\forall x)(Mx) \vee (\forall y)(Py)]$ by IF
     $(\exists x)(\sim Mx) \vee (\forall y)(Py)$ by QN
PREM 2: $(\exists x)[Px \to (\forall y)(Sy)]$
Similarly,
     $(\exists x)(\sim Px) \vee (\forall y)(Sy)$
CON: $(\exists x)[Mx \to (\forall y)(Sy)]$
Similarly,
     $(\exists x)(\sim Mx) \vee (\forall y)(Sy)$

~CON: $(\forall x)(Mx) \land (\exists y)(\sim Sy)]$ by NOR, QN
~CON A: $(\forall x)(Mx)$ by SIMP
~CON B: $(\exists y)(\sim Sy)$ by SIMP

We obtain much the same truth tree:

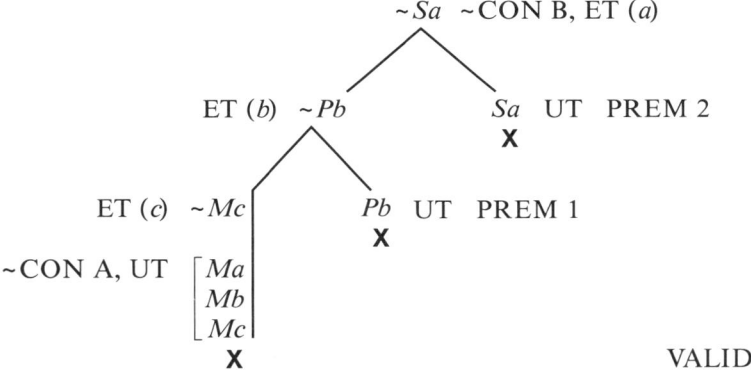

VALID

Accordingly, the truth-tree method, like deduction, establishes that argument L is valid.

Some Misleading Arguments

Finally, before we go on to deal with the logic of relations, I want to discuss briefly two types of argument that tend to mislead.

M. Such arguments as the following do not look valid, but they are.

All fishermen are generous.
No judges are generous.
Some judges are fishermen.
Therefore, some housewives are generous.

PREM 1: $(\forall x)(Fx \to Gx)$
PREM 2: $(\forall x)(Jx \to \sim Gx)$
PREM 3: $(\exists x)(Jx \wedge Fx)$
CON: $(\exists x)(Hx \wedge Gx)$
~CON: $(\forall x)(Hx \to \sim Gx)$ by QN'

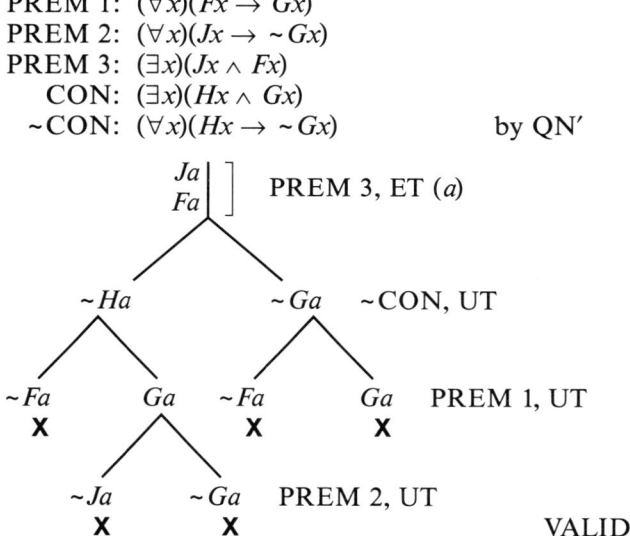

The argument seemed invalid, because the conclusion seemed unrelated to the premises. But the tree closes. Perhaps we can see more clearly why the tree closes if we pursue that intuition of "unrelatedness" and redo the truth tree, changing the order of work:

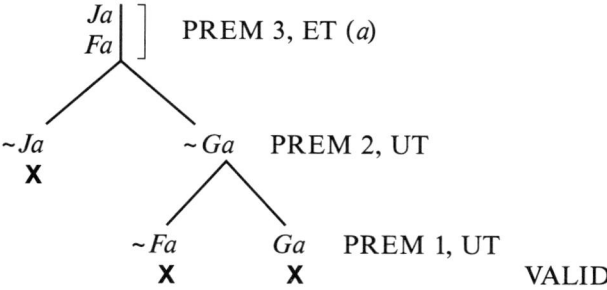

The tree closes, even before the conclusion is entered.

Truth Trees for Predicate Logic

This tree shows that the premises were inconsistent. From such premises, any conclusion can be derived, either by reductio or by some related device. But such premises, being inconsistent, cannot hold; so such arguments are never straightforwardly useful. On the other hand, they are harmless; since we cannot accept their premises, the arguments cannot mislead us as to how things are. Arguments with inconsistent premises may seem not to be valid, but they are.

N. Such arguments as the following may look valid, but they are not.

All Spartans are brave.
All Spartans are Greeks
Therefore, some Greeks are brave.

PREM 1: $(\forall x)(Sx \to Bx)$
PREM 2: $(\forall x)(Sx \to Gx)$
   CON: $(\exists x)(Gx \wedge Bx)$
 ~CON: $(\forall x)(Gx \to \sim Bx)$ by QN'

```
                  ~Ga              ~Ba        ~CON, UT

          ~Sa          Ga    ~Sa         Ga    PREM 2, UT
                       X

      ~Sa    Ba    ~Sa   Ba  ~Sa   Ba          PREM 1, UT
       ↑      ↑      ↑    X    ↑    X
                                               INVALID
```

The reader will notice that every open branch on this tree contains '$\sim Sa$'. What made the argument seem valid at first was the natural assumption that there are Spartans. If we add "There are Spartans" explicitly to the original premises, we obtain a valid argument, N':

N'. PREM 1: $(\forall x)(Sx \to Bx)$
PREM 2: $(\forall x)(Sx \to Gx)$
PREM 3: $(\exists x)(Sx)$
CON: $(\exists x)(Gx \land Bx)$
~CON: $(\forall x)(Gx \to \sim Bx)$ by QN'

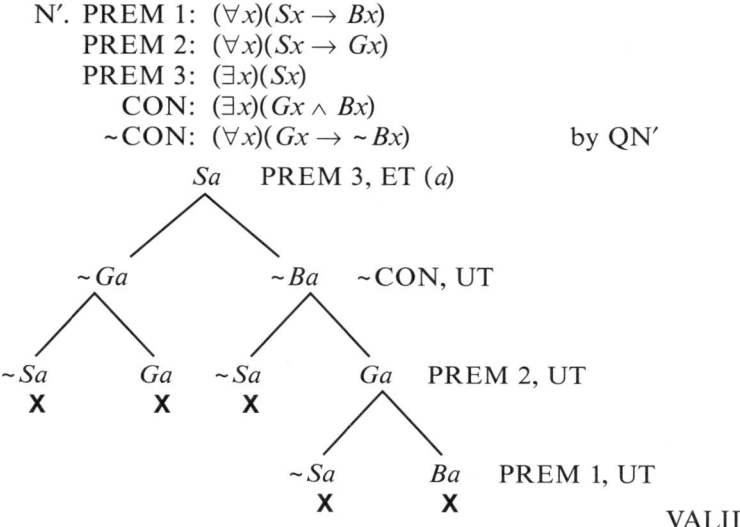

VALID

The problem exemplified by this pair of arguments is the so-called "problem of the existential import of universals." Modern notation, which writes "All As are Bs" as '$(\forall x)(Ax \to Bx)$', marks a departure from the Aristotelian tradition; for Aristotle thought that "All As are Bs" did have "existential import", that it conveyed the information that there exist some As. '$(\forall x)(Ax \to Bx)$' does not.

Aristotle, who was a biologist as well as a logician and philosopher, thought it would make no sense to claim that all frogs croak if there were no frogs, or to claim that all fish have gills if there were no fishes. Modern logicians, who want their notation to be useful for physics and ethics and the law, as well as for biology, prefer not to make such existential assumptions. Claims about perfect gases and perfect men and women make good sense, whether or not there are any perfect gases or perfect men and women. And when we promulgate laws and say, for example, that "All trespassers will be punished", we hope that there will be no trespassers.

The concerns of the biologist can be accommodated by explicit statement of existential claims where they are needed – and justified. There are contexts, therefore, in which a case can be made for symbolizing such statements as "All Spartans are brave" like this:

$(\forall x)(Sx \to Bx) \land (\exists x)(Sx)$

explicitly stating that the subject term is nonempty. But the case must be made, and the existential claim must be explicit. For our purposes, argument N' is valid, and argument N is not.

One further point about the interpretation of universal claims. Such sentences as "All men are not agreeable" in English are radically ambiguous. They can be read as universals:

$(\forall x)(Mx \to {\sim} Ax)$
${\sim}(\exists x)(Mx \land Ax)$

All men are disagreeable.
No men are agreeable.

or, alternatively, as the denials of universals:

${\sim}(\forall x)(Mx \to Ax)$
$(\exists x)(Mx \land {\sim} Ax)$

Not all men are agreeable.
Some men are disagreeable.

The only way I know to avoid this ambiguity is to avoid the English locution "All A are not B" altogether (which is what I try to do). In context, one must make a good guess as to what the speaker means and hold her to it.

Reminders for Chapter 5

ET: In place of an existentially quantified sentence or component of a sentence, to be entered on an open path of a tree, enter an instance of it; in each case enter a *new* name.

UT: In place of a universally quantified sentence or component of a sentence, to be entered on an open branch of a truth tree, enter *every* instance of it that uses a name or pseudoname already appearing on the branch in question.

Where possible, enter "real names" – names that occur in the premises or conclusions of the argument being tested – *before* use of ET or UT.

Where there is a choice, use ET before UT.

In every case, *internalize negation* (use QN or QN') before building a truth tree.

QN $\sim(\forall x)(Fx) \leftrightarrow (\exists x)(\sim Fx)$
     $\sim(\exists x)(Fx) \leftrightarrow (\forall x)(\sim Fx)$

QN' $\sim(\forall x)(Ax \to Bx) \leftrightarrow (\exists x)(Ax \wedge \sim Bx)$
     $\sim(\exists x)(Ax \wedge Bx) \leftrightarrow (\forall x)(Ax \to \sim Bx)$

For rules of inference, see Reminders for Chapter 4.

Problems for Chapter 5

Parts I and II of these problems depend only on the initial explanation of truth trees for predicate logic (pages 187–200) and the comment on argument N (pages 217–219). The other problems are best left until the whole chapter has been read.

I. Consider the following arguments, many of them adapted from Lewis Carroll. In each case, state whether the argument seems to be valid. Check this judgment by truth tree. If the argument is valid, do a formal deduction of its conclusion from its premises. If the argument is invalid, explain why.

   1. Every eagle can fly.
      Some pigs cannot fly.
      Therefore, some pigs are not eagles.

   2. Every eagle can fly.
      Some pigs can fly.
      Therefore, some pigs are eagles.

3. All who are anxious to learn work hard.
   Some of these boys work hard.
   Therefore, some of these boys are anxious to learn.

4. All lions are fierce.
   Some lions do not drink coffee.
   Therefore, some coffee drinkers are not fierce.

5. All lions are fierce creatures.
   Some lions do not drink coffee.
   Therefore, some fierce creatures do not drink coffee.

6. No school principals are mischievous.
   Some monkeys are mischievous.
   Therefore, some monkeys are not school principals.

7. No misers are unselfish.
   None but misers save egg-shells.
   Therefore, no one who is unselfish saves egg-shells.

8. All frogs are unpoetical.
   Some ducks are poetical.
   Therefore, some frogs are not ducks.

9. Residents who are citizens are eligible to vote.
   Only citizens are residents.
   Therefore, all residents are eligible to vote.

10. No engineers will be hired who are neither well trained nor experienced.
    There are no experienced engineers who are women.
    Some women engineers are being hired.
    Therefore, some women engineers are well trained.

11. All pillows are soft.
    No pokers are soft.
    Therefore, some pillows are not pokers.

12. Wine drinkers are very communicative.
    One can always trust a very communicative person.
    Therefore, wine drinkers are trustworthy.

13. Dentists are dreaded by children.
    No emperors are dentists.
    Therefore, emperors are not dreaded by children.
14. All my uncles are either generous or very amusing.
    Those of my uncles who are generous are also amusing.
    Therefore, all my uncles are amusing.
15. Only philosophy majors who have taken logic may join this seminar.
    Some philosophy majors have not taken logic.
    Therefore, some philosophy majors may not join the seminar.
16. No basketball players play chess.
    Some mathematicians play chess and some do not.
    Therefore, some mathematicians play neither chess nor basketball.
17. All basketball players play chess.
    Some mathematicians play chess and some do not.
    Therefore, some mathematicians play neither chess nor basketball.
18. Prudent travellers carry plenty of small change.
    Imprudent travellers lose their baggage, and some of them smoke cigars.
    Therefore, some cigar smokers do not carry plenty of small change.
19. No pokers are soft.
    Some pillows are soft, and some are not.
    Therefore, not everything is a poker.
20. All chess players have a combative streak.
    Everyone who is well informed and combative makes a good debater.
    Therefore, well-informed chess players make good debaters.

21. Imprudent travellers lose their luggage and some of them smoke cigars.
    All bankers who travel are very prudent.
    Therefore, some who lose their luggage are not bankers.
22. Bankers who travel are very prudent.
    Imprudent travellers lose their luggage.
    Therefore, some who lose their luggage are not bankers.
23. To join this club one must have either roller skates or a bicycle.
    All skaters in the club have bicycles.
    Therefore, everyone in the club has a bicycle.
24. My oldest friends are women who went to school with me.
    Not all who went to school with me are women.
    Therefore, some who went to school with me are not among my oldest friends.
25. No one I know, except for my relatives, plays the cello.
    Some cellists I know also play the piano.
    Therefore, some of my relatives do not play the piano.
26. No one I know, except for my relatives, plays the piano.
    Some cellists I know also play the piano.
    Therefore, some cellists are relatives of mine.

II. Assess the validity of each of the following *statements*, and check your judgment by truth tree. If the statement is valid, do a deduction to establish it. If it is not valid, explain why not.

1. It is not the case both that only athletes enjoy this game and that everyone who enjoys this game is unathletic.
2. Either wine drinkers are very communicative or wine drinkers are not very communicative.
3. If all the politicians of my acquaintance are honest and all the politicians of my acquaintance are dishonest then I am not acquainted with any politicians.

4. Some of those who attended the concert last night were enthusiastic about it or else were not enthusiastic about it.

5. Everyone who was enthusiastic about last night's concert either knew what he was talking about or did not.

III. Carry out the deduction, sketched in the text, for the argument:

All freshmen have computers.
If no one who has a computer can spell, then everyone who has a computer uses a spelling checker.
Therefore, if any freshmen do not use spelling checkers, some computer owners can spell.

IV. Use truth trees to establish the following four equivalences, which have already been established by deduction; your work will be clearer if, in each case, you build two truth trees, one to establish the left-to-right implication, the other to establish the implication from right to left. Mark your trees clearly.

1. $[(\forall x)(Fx) \to p] \leftrightarrow (\exists x)(Fx \to p)$ (rule of passage 8)
2. $[p \lor (\forall x)(Fx)] \leftrightarrow (\forall x)(p \lor Fx)$ (rule of passage 3)
3. $[(\exists x)(Fx) \lor (\exists x)(Gx)] \leftrightarrow (\exists x)(Fx \lor Gx)$
4. $[(\forall x)(Fx) \land (\forall x)(Gx)] \leftrightarrow (\forall x)(Fx \land Gx)$

V. Assess the validity of each of the following arguments, and check your judgment by truth tree. If the argument is valid, deduce its conclusion formally from its premise; if it is not valid explain why.

1. If anyone takes the swimming test and passes, Joe will pass it if he takes it.
Therefore, if Joe takes the test and passes, everyone who takes it will pass.

2. If anyone takes the test and passes it, Joe will pass it if he takes it.
Therefore, if Joe takes the test but does not pass, no one who takes it will pass.

3. If anyone takes the test and passes it, Joe will pass it if he takes it.
   If Joe and Henry both take the test, Joe will pass if Henry does.
4. If everyone is considerate, everyone benefits.
   Therefore, everyone who is considerate benefits.
5. If someone is considerate, everyone benefits.
   Therefore, everyone who is considerate benefits.
6. If someone who witnessed the crime either told the truth or did not, someone will be apprehended.
   Therefore, someone will be apprehended.
7. All freshmen have computers.
   If no computer owners can spell, everyone who has a computer uses a spelling checker.
   Some freshmen do not use spelling checkers.
   Therefore, some freshmen can spell.
8. If the newspaper reports are correct, the warden is to be believed and none of the escaped prisoners have guns.
   If the warden is to be believed, not all of the escaped prisoners are dangerous.
   Therefore, if the newspaper reports are correct, not all people with guns are dangerous.
9. If the newspaper reports are correct, the warden is to be believed and none of the escaped prisoners have guns.
   If the warden is to be believed, some of the escaped prisoners are dangerous.
   Therefore, if the newspaper reports are correct, not all dangerous people have guns.
10. If no students majoring in art history take economics, some students majoring in art history will take the history of philosophy.
    Therefore, some students major in art history.
11. If some students either pay their bills or do not pay their bills, some students will not graduate.
    Therefore, some students will not graduate.

12. If all the students either pay their bills or do not pay their bills, some students will graduate.
    Therefore, some students will graduate.
13. If anyone has a gun, all children need to be protected.
    Therefore, if some children have guns, everyone needs to be protected.
14. If anyone has a gun, all children need to be protected.
    Therefore, if everyone has a gun, all children need to be protected.
15. If anyone has a gun, all children need to be protected.
    Therefore, some children need to be protected if someone has a gun.
16. If anyone has a gun, all children need to be protected.
    Therefore, if some children have guns, some children need to be protected.
17. If any teenagers have guns, all teenagers need to be protected.
    Therefore, if all teenagers have guns, all teenagers need to be protected.
18. If any teenagers have guns, all children need to be protected.
    Therefore, if all teenagers have guns, all children need to be protected.
19. If some students in this class do very well in science, some students in this class will get scholarships.
    Therefore, if some students in the class do very well in science, they will get college scholarships.
20. All students in this class who do well in science get college scholarships.
    Therefore, if some students in this class do well in science, some students in this class will get college scholarships.

Part III
Relational Logic

Chapter 6
New Restrictions on the Rules of Inference

We turn next to quantification of somewhat greater complexity than that discussed in Part II: here the open sentences will have more than one variable, and, as a matter of routine, the quantifiers will be nested; that is, quantifiers will occur within the scopes of other quantifiers. Such quantification is needed in order to deal with relationships.

The need for a relational logic had been recognized long before a notation convenient for developing it was devised. The ancient Greeks knew, for example, that such arguments as the following are valid:

(1) Someone is a mother.
   Therefore, somebody has a mother.
(2) All circles are plane figures.
   Therefore, everyone who draws a circle draws a plane figure.
(3) There is someone whom everybody loves.
   Therefore, everybody loves someone.

Neither the Aristotelian theory of the syllogism, however, which is adequate for most of the predicate logic we covered in Part II, nor the propositional logic of the Stoics, which dealt with much of what we covered in Part I, could make clear why such arguments as these are valid. We shall use the principles of inference and the notation of chapter 4, extended only to allow more than one variable in each open sentence, to establish the validity of each of the three arguments above. Further restrictions on the rules of inference will also be needed, but we shall defer explanation of those restrictions until after the initial discussion of these arguments.

(1) Someone is a mother.
Therefore, somebody has a mother.[1]

We need only one definition:

Let $Mxy = x$ is mother to y.

This definition tells us that

$Mab = a$ is mother to b.

and

$Mcd = c$ is mother to d.

etc.

Notice that 'Mab' does *not* stand for "b is mother to a"; the order matters. We could have set up the definition the other way around: "Let $Mxy = y$ is mother to x" – but we didn't. Once the definition has been stipulated, it must be respected. And, because order in definition matters, we must be careful always to be explicit about it and to define our terms so as to make clear how many variables are involved and in what order. Let us note also that

Let $Myx = y$ is mother to x.

or

Let $Mwz = w$ is mother to z.

is the *same* as the definition given above; any one of them could be written:

Let $M\_\_ \ldots = \_\_$ is mother to ...

Now, how do we write the claim that Alice is "a mother"? *Not* 'Ma'. In this context 'Ma' tells us nothing. It is ungrammatical; for 'M' has been defined as a two-place predicate, a relation, not

---

[1] The reader will recognize that this holds for uncles and supervisors, as well as mothers, but not for engineers or Frenchmen, in other words, that this is, basically, a relational argument. Notice also that, in order to make reading a little easier, I have used 'someone' for the mother and 'somebody' for the child. There is no difference in meaning between these words; they only *suggest* that they refer to different entities, and help us to keep track of which is which.

as a one-place predicate, and the expression 'Ma' is undefined; in this relational context it does not work. We find ourselves in need of an "auxiliary" definition.

To say that Alice is a mother is to say that Alice has a child: there is someone that Alice is mother to; i.e.,

$$(\exists y)(May)$$

or, if you prefer, $(\exists x)(Max)$; the choice of bound variable is indifferent. What matters is that Alice appears in the first place and the variable of quantification appears in the second.

Similarly, we paraphrase the statement that Bertrand has a mother in relational terms: there is someone who is mother to Bertrand, that is,

$$(\exists x)(Mxb) = b \text{ has a mother.}$$

Accordingly, we have:

$$(\exists y)(Mxy) = x \text{ is a mother.}$$

and

$$(\exists x)(Mxy) = y \text{ has a mother.}$$

Notice that, in both cases, the formulas defined, which are both open sentences, each of them containing one free variable, characterize the entity marked by the free variable, the variable not bound by the quantifier. The first formula tells us about x that x is a mother; the second tells us about y that y has a mother. The argument we are formalizing becomes

(1) $(\exists x)[(\exists y)(Mxy)]$ [There is someone (x) who is mother to somebody (y).]
∴ $(\exists y)[(\exists x)(Mxy)]$ [There is somebody (y) that someone (x) is mother to.]

In the premise the quantifier '$(\exists x)$' has wide scope: its scope is the quantified but still open sentence '$(\exists y)(Mxy)$', or "x is a mother". In the conclusion the same quantifier has narrow scope: 'Mxy', or "x is mother to y". Vice versa for '$(\exists y)$'. The brackets help to make this nesting obvious. But they are not really needed, for the formulas are quite unambiguous without them, and we shall ordinarily omit them, once the reader has become accustomed to nested quantifiers.

Here is a deduction:

* 1	$(\exists x)[(\exists y)(Mxy)]$	PREM
* 2	$(\exists y)(May)$	1 EI (a)
* 3	Mab	2 EI (b)
* 4	$(\exists x)(Mxb)$	3 EG
* 5	$(\exists y)[(\exists x)(Mxy)]$	4 EG
		QED

Note that, with each use of EI, the outside quantifier – and only the outside quantifier – is dropped, and that each use of EG introduces an outside quantifier, covering the whole formula.

Note also that no two quantificational steps may be collapsed into one. This is a matter of some importance. It is a mistake not to make quite clear which step is taken first. Reading the deduction above in ordinary language, we have:

* 1 Someone is a mother. That is,
    There is someone who is mother to somebody. PREM

* 2 Let's call her Alice.
    Alice is mother to somebody.
    That is,
    There is somebody that Alice is mother to. 1 EI (a)

* 3 Let's call him Bertrand.
    Alice is mother to Bertrand.
    Alice is Bertrand's mother. 2 EI (b)

* 4 There is someone who is Bertrand's mother.
    That is,
    Bertrand has a mother. 3 EG

* 5 There is somebody who has a mother.
    That is,
    Somebody has a mother. 4 EG
    QED

Step by step, this little deduction makes sense. But it would not make sense to move directly from the premise to step 4, dropping the middle quantifier '(∃y)', instead of the outside quantifier '(∃x)'. That inference would read something like this:

* 1 Someone is a mother. That is,
 Someone has a child. PREM
* 2' Let's call him Barry. Barry has a mother. 1 EI (b)

This does not feel like a reasonable *direct* inference. It follows, of course, as we have seen, but it is *not* an instance of EI.

Note finally that either of the following two symbolic variants of steps 4 and 5 would be acceptable, although the version given above is for most readers easiest to follow:

* 4' (∃z)(Mzb) 3 EG
* 5' (∃w)[(∃z)(Mzw)] 4' EG
  QED

* 4" (∃y)(Myb) 3 EG
* 5" (∃x)[(∃y)(Myx)] 4" EG
  QED

The advantage of the first version of steps 4 and 5 is that the open sentences in steps 1 and 5 are exactly alike: 'x' is proxy for the mother; 'y' is proxy for the child. This sort of uniformity – where we can manage it – makes for ease in reading. It mimics the way we use pronouns in ordinary language. But it is not essential. The reader may have noticed that, in my ordinary-language version of this argument I used 'someone' and 'somebody', 'her' and 'him', similarly, without any significant reference to gender – the argument, after all, would have worked out the same way if we had been talking about fathers or about parents or about supervisors.

We turn next to the second example:

(2) All circles are plane figures.
 Therefore, everyone who draws a circle draws a plane figure.

This presents a new problem in formalization. We know how to formalize "All circles are plane figures": '$(\forall x)(Cx \to Px)$' or

'$(\forall y)(Cy \to Py)$'. But how do we handle "Everyone who draws a circle draws a plane figure"?

Let $Cx = x$ is a circle.
Let $Px = x$ is a plane figure.
Let $Dxy = x$ draws y.

Again we break down the problem (that of formalizing the conclusion) into smaller units and provide auxiliary definitions. The conclusion tells us that, no matter whom we pick, if he draws a circle he draws a plane figure:

$(\forall x)(x$ draws a circle $\to x$ draws a plane figure$)$

$(\forall x)(\_\_ x \_\_ \to \ldots x \ldots)$

How do we formalize the antecedent, "x draws a circle"? We observe that what that open sentence says about x is that there is a circle that x draws:

$(\exists y)(Cy \land Dxy)$

Similarly, the consequent, "x draws a plane figure", becomes

$(\exists y)(Py \land Dxy)$

or, if we prefer,

$(\exists z)(Pz \land Dxz)$

so the conclusion becomes

$(\forall x)[(\exists y)(Cy \land Dxy) \to (\exists y)(Py \land Dxy)]$

or

$(\forall x)[(\exists y)(Cy \land Dxy) \to (\exists z)(Pz \land Dxz)]$

It does not matter here whether or not the variable of quantification in the antecedent of the open sentence matches the variable of quantification in the consequent; such matching would be of no consequence, since the scopes do not overlap. Nor does the original ordinary-language sentence address the question of whether the circle that one draws is the same item as the plane figure that one draws. But there must be and there is cross reference between the 'x' of the antecedent and the 'x' of the consequent – these are both bound by the outside quantifier '$(\forall x)$'.

Once the conclusion has been correctly formalized, the deduction is not hard to find. Given the premise, we assume that Alfred draws a circle (step 2 below). It is worth noting that the auxiliary premise we take here is *not* "Someone draws a circle," but rather "Alfred draws a circle," in view of the strategy of proof we plan to follow – a strategy familiar from chapter 4. We will need an *unflagged* letter *a* so that we can generalize at the end of the deduction (see step 10 and step 11 below). We show that, since Alfred draws a circle and all circles are plane figures, Alfred draws a plane figure (step 9 below); then we use Conditional Proof and UG to complete the deduction.

To Prove: $(\forall x)[(\exists y)(Cy \wedge Dxy) \rightarrow (\exists z)(Pz \wedge Dxz)]$

*	1	$(\forall x)(Cx \rightarrow Px)$	PREM
**	2	$(\exists y)(Cy \wedge Day)$	PREM
**	3	$Cb \wedge Dab$	2 EI (*b*)
**	4	Cb	3 SIMP
**	5	Dab	3 SIMP
**	6	$Cb \rightarrow Pb$	1 UI
**	7	Pb	6,4 MP
**	8	$Pb \wedge Dab$	7,5 ADJ
**	9	$(\exists z)(Pz \wedge Daz)$	8 EG
*	10	$(\exists y)(Cy \wedge Day) \rightarrow (\exists z)(Pz \wedge Daz)$	2–9 CP
*	11	$(\forall x)[(\exists y)(Cy \wedge Dxy) \rightarrow (\exists z)(Pz \wedge Dxz)]$	10 UG (*a*)
			QED

We check for double flagging; there is none. This a valid argument. But the checking procedure is not yet complete. I will, however, continue to postpone discussion of the new restrictions that will be needed for EI and UG (which will necessitate further checking) until after we have looked at the last of our three arguments.

(3) There is someone whom everybody loves.
Therefore, everybody loves someone.

Let $Lxy = x$ loves y

$(\exists x)[(\forall y)(Lyx)]$
∴ $(\forall y)[(\exists x)(Lyx)]$

or, more succinctly,

$$(\exists x)(\forall y)(Lyx)$$
$$\therefore (\forall y)(\exists x)(Lyx)$$

Before we go on to provide a deduction, I want to make two points in regard to the ordinary-language version. First, as given, the conclusion, "Everybody loves someone," is ambiguous: it can be interpreted as I have interpreted it, but it can also be read as a mere rephrasing of the premise. This misreading can be avoided by expanding slightly:

> Therefore, everybody loves someone or other.

which yields the formal version of the conclusion given above, without question. But one should not look for anything to correspond to "or other" in the formal version – say, '$\vee Lyz$'. For the phrase "or other" adds nothing. It serves only to remove ambiguity.

Second, the argument as given is immediately recognizable as valid: if there is someone whom everybody loves, say Joe, then each of us loves Joe, and so each of us loves someone.

But the converse:

> Each of us loves someone or other.
> Therefore, there is someone whom everybody loves.

is just as immediately recognizable as *invalid*. Given the premise, we could love each other in pairs or in groups, or half of us could love one person and the other half love someone else, and so forth – with no reason even to suggest that there is any one person who is loved by everybody.

This leads to the observation – to be supported later both by deduction and by testing procedures – that the order of nested quantifiers determines relations of implication. Where the open sentences covered are alike, a sentence of the form $(\exists x)(\forall y)(\_\_)$ will imply a sentence of the form $(\forall y)(\exists x)(\_\_)$, but not vice versa. $\exists \forall$ implies $\forall \exists$ (in the spoken language, EA implies AE), but not vice versa. This observation is often useful, but it is reliably useful only when the sentences have been very carefully formalized.

New Restrictions on the Rules of Inference

It is easy to confuse ∃∀ with ∀∃, and it is easy to make mistakes in formalizing the open sentences as well.

Here is a deduction for the valid one of our pair of arguments about love:

* 1	(∃x)(∀y)(Lyx)	PREM
* 2	(∀y)(Lya)	1 EI (a)
* 3	Lba	2 UI
* 4	(∃x)(Lbx)	3 EG
* 5	(∀y)(∃x)(Lyx)	4 UG (b)
		QED

In ordinary language,

* 1 There is someone whom everybody loves.   PREM

    To Prove: Everybody loves someone or other.

* 2 Let's call him Algernon. Everybody loves Algernon.   1 EI (a)

* 3 So Barbara, for example, loves Algernon.   2 UI

* 4 There is someone (Algernon) whom Barbara loves.   3 EG
    That is, Barbara loves someone.

* 5 But Barbara was just anybody (see step 3); so everybody loves someone or other.   4 UG (b)
    QED

Cross Flagging

And here is a deduction for the invalid one:

* 1	(∀y)(∃x)(Lyx)	PREM
* 2	(∃x)(Lax)	1 UI
* 3	Lab	2 EI (b)
* 4	(∀y)(Lyb)	3 UG (a)
* 5	(∃x)(∀y)(Lyx)	4 EG

What has gone wrong? The deduction is "finished": neither 'a' nor 'b' occurs either in the premise or in the conclusion. There is no double flagging. Nonetheless, this little derivation has "established" what is obviously an invalid argument; so there must be something wrong with it. Let us read the steps in English.

* 1 Everyone loves someone or other.   PREM

* 2 So Albert, for example, loves someone or other.   1 UI

* 3 Let's call her Beatrice; Albert loves Beatrice.   2 EI (b)

* 4 But Albert (see step 2) was just anybody; so, since Albert loves Beatrice, everybody loves Beatrice.   3 UG (a)

* 5 There is someone (Beatrice) whom everybody loves.   4 EG

Step 4 makes no sense. Remember that the restrictions on UG discussed in chapter 4 were designed to block generalization from a special case.

When Albert was introduced at step 2, Albert was just anyone; there was nothing special about Albert. So whatever could be shown to be true of Albert at that point would be shown thereby to be true of everyone – say, if we showed that Albert loves someone or owns a tennis racket, it would follow that everyone loves someone or owns a tennis racket. But the situation is different at step 4, *in view of step 3*. At step 3, specificity is introduced. Beatrice is not just anyone; she has been chosen from the possibly very small class of people whom Albert loves, and this specificity is registered by the flag '(b)'. What has been shown to be true of Albert *in relation to Beatrice* may very well not be true of everyone.

At step 4, if we were to generalize from b (Beatrice, instead of Albert), as follows:

* 3 Lab 2 EI (b)
* 4' $(\forall x)(Lax)$ (Albert loves everybody.)   3 UG (b)

this would have been obviously fallacious, by the fallacy of double flagging. But our step 4 is equally fallacious; for, at step 3, Albert has lost his generality and become, like Beatrice, a special case – at least in relation to Beatrice. Beatrice's specificity has rubbed off on Albert. The alternative step 4' fails because b is a special case, flagged at step 3. Our step 4 fails because a has become a special case, infected by the specificity of b.

Accordingly, in order to block such fallacious reasoning, we articulate a third restriction[2] on EI and UG, and prohibit *cross flagging*:

1. A deduction in which any flag occurs must be finished; that is, no flagged letter may occur either in its conclusion or in any of its undischarged premises.

2. Within a deduction no letter may be flagged twice. Using the same flag twice is the *fallacy of double flagging*.

3. Within a deduction, if a pseudoname occurs at a step at which another pseudoname is flagged, the first pseudoname may not be flagged at any step at which the other pseudoname occurs.

    Violation of this restriction is the *fallacy of cross flagging*.

In order to make it easier to check our deductions for cross flagging, we extend the flagging system, to mark, at each flagged step, the pseudonames that occur there, and we use square brackets for that purpose. This is how that little fallacious deduction will look:

```
* 1  (∀y)(∃x)(Lyx)           PREM
* 2  (∃x)(Lax)               1 UI
* 3  Lab                     2 EI   (b) [a]
                                       ╳
* 4  (∀y)(Lyb)               3 UG   (a) [b]
* 5  (∃x)(∀y)(Lyx)           4 EG
```

---

[2] The first two restrictions are repeated from Chapter 4. See also Reminders for Chapter 4. What is new is restriction 3.

At step 3, *a* occurs where *b* is flagged; at step 4, *b* occurs where *a* is flagged. The pattern of letters on the right exhibits cross flagging in a graphic way.

The reader should now look back at the three deductions we have shown in this chapter and check explicitly for cross flagging.

Finally we note a fourth restriction on EI and UG: *cyclical flagging* is likewise prohibited. If *a* occurs where *b* is flagged, and *b* occurs where *c* is flagged, out to some pseudoname *n*, *n* may not occur where *a* is flagged. This is just an extension of the notion of cross flagging.

In summary, a deduction in which EI or UG is used will be valid only if the deduction (1) is finished, and contains (2) no double flagging, (3) no cross flagging, and (4) no cyclical flagging.

Another, related, argument is of some interest. Although this argument is valid, its validity is for many students counterintuitive, and it may be worth while to try to understand why.

All the world loves a lover.[3]
Someone loves somebody.
Therefore, everyone loves everybody.

To begin with, the argument depends upon a rather odd definition of "a lover" as someone who loves somebody, in analogy with the conventional definition of "a mother" which we have been using, as someone who is mother to somebody. One tends to think of a lover as a romantic young fellow like Shakespeare's Romeo or, perhaps, as a long-term live-in companion, not as just someone who loves somebody or other. But this minimal definition is not, I

---

[3] This is a familiar saying. Bartlett [John Bartlett, *Familiar Quotations* (Boston: Little, Brown & Co., 13th ed., 1955), p. 411] attributes it to Ralph Waldo Emerson: "All mankind love a lover" in Emerson's essay "Love". See *Emerson's Essays* (London: J.M. Dent & Sons, Everyman edition, 1971), p. 100; "Love" is Essay #5 in the first series, originally published in 1841. In the twentieth-century literature on logic this saying has come to belong to Richard Jeffrey; see *Formal Logic*, p. 145.

think, what makes the argument surprising. If we rephrase the first premise along the lines of that definition, say, as "Anyone who loves somebody or other is loved by everyone," the argument is still surprising. A sketch of a deduction may be helpful.

Let us suppose that all the world does love a lover, that is, that anyone who loves somebody is loved by all of us (Premise 1). And let us suppose further (Premise 2) that someone does love somebody – say, Romeo loves Juliet (by two steps of EI). Since Romeo loves somebody, if we apply the first premise we will find that all of us love Romeo.

We all love Romeo. But do we therefore love everybody? To claim at this point that we do would be to argue fallaciously, to generalize from a special case. We notice next that Romeo loves Juliet. We all love Romeo, but do we all, like Romeo, love Juliet? To claim that we do would be, again, to generalize from a special case. Whichever way we try to generalize at this point, we are blocked by the fallacy of double flagging. It is this recognition, I think, that prompts one to question the validity of the argument.

But let us persist. The sketch so far has led to an impasse. To reach the conclusion "Everyone loves everybody", it will be necessary to generalize. But we cannot generalize from Romeo (or Juliet). Can we generalize from someone else? Have we shown about anyone else that he/she is a lover? Or a loved one? Of course we have. All of us are lovers, for we all love Romeo. So let us pick any one of us we please, for example, Pat. Pat is a lover, since he/she loves Romeo. The first premise tells us, accordingly, that all of us love Pat. *Now* we can generalize. Everyone loves Pat; Pat was just anybody, so everyone loves everybody. QED.

Now let us do some of the formal work. As Professor Jeffrey indicates, the first premise can be formalized in a number of equivalent ways. Here are some of them:

$$(\forall x)[(\exists y)(Lxy) \rightarrow (\forall z)(Lzx)]$$

$$(\forall x)(\forall y)[Lxy \rightarrow (\forall z)(Lzx)]$$

$$(\forall x)(\forall y)(\forall z)(Lxy \rightarrow Lzx)$$

$$(\forall x)(\forall z)[(\exists y)(Lxy) \rightarrow Lzx]$$

The reader should verify that all of these (and more) can be shown to be equivalent, using the rules of passage. Because the first of these is the easiest to read – and, in this case, convenient for deduction – we use the first one.

$*\begin{cases} 1 & (\forall x)[(\exists y)(Lxy) \to (\forall z)(Lzx)] \\ 2 & (\exists x)(\exists y)(Lxy) \end{cases}$ PREM

To Prove: $(\forall x)(\forall y)(Lxy)$

* 3 $(\exists y)(Lry)$ 2 EI (r)
* 4 $(\exists y)(Lry) \to (\forall z)(Lzr)$ 1 UI
* 5 $(\forall z)(Lzr)$ 4, 3 MP
* 6 Lpr 5 UI
* 7 $(\exists y)(Lpy)$ 6 EG
* 8 $(\exists y)(Lpy) \to (\forall z)(Lzp)$ 1 UI
* 9 $(\forall z)(Lzp)$ 8, 7 MP
* 10 Lap 9 UI
* 11 $(\forall y)(Lay)$ 10 UG (p)
* 12 $(\forall x)(\forall y)(Lxy)$ 11 UG (a)

                                          QED

The reader will remember that the "sticking point" in this deduction came after step 5 (Everyone loves Romeo), where one might be tempted to generalize from Romeo, fallaciously, and where we were forced to take a different tack. When we get to step 9 (Everyone loves Pat), by contrast, generalizing would be OK. The result might be

* 10′ $(\forall x)(\forall z)(Lzx)$ 9 UG (p)

which is equivalent, by CV, to

* 10″ $(\forall x)(\forall y)(Lyx)$ 10 CV

in English,
          Everyone is loved by everybody.

But we set out to prove
          $(\forall x)(\forall y)(Lxy)$
          Everyone loves everybody.

In order to prove precisely what we set out to prove, those last three steps were needed.

Alternatively, we could have used UG at step 10 and gone from there to derive the desired conclusion. This would suggest that '$(\forall x)(\forall y)(Lxy)$' and '$(\forall y)(\forall x)(Lxy)$' – reversing the universal quantifiers – are interchangeable, as indeed they are. Similarly, the argument (1) about the mothers, which we discussed at the beginning of this chapter, suggests that '$(\exists x)(\exists y)(Mxy)$' and '$(\exists y)(\exists x)(Mxy)$' are interchangeable. After all, the converse of (1):

Somebody has a mother.
Therefore, someone is a mother.

$$\frac{(\exists y)(\exists x)(Mxy)}{\therefore \ (\exists x)(\exists y)(Mxy)}$$

is obviously valid and similarly provable.

In other words, a pair of adjacent existential quantifiers is reversible, and, likewise, a pair of adjacent universal quantifiers is reversible. The following equivalences hold:

$(\exists x)(\exists y)(Fxy) \leftrightarrow (\exists y)(\exists x)(Fxy)$
$(\forall x)(\forall y)(Fxy) \leftrightarrow (\forall y)(\forall x)(Fxy)$

The reader should satisfy himself that this is so, by spelling out the relevant deductions, and should feel free thereafter to use these two equivalences as needed.

The Scope of Quantifiers

We have noted (in chapter 4) that correct formalization of ordinary-language sentences requires careful attention to the scope of quantifiers. In relational logic variables proliferate and with them opportunities for confusion, in particular for failure of cross reference. As we all know, the natural languages we use are rich in pronouns and other linguistic devices that allow us to cross-refer beyond the narrow contexts in which our pronouns are introduced; as a result, a pronoun can hold its reference almost indefinitely, depending on the skill of the speaker and the memory of the listener. In contrast, in quantifier notation a bound variable holds its identity only within the scope of the quantifier that binds it, and cross reference is strictly limited. This limitation leads to problems in formalization.

Let us look at some examples.

(1) If something is broken, someone will pay for it.
(2) Everyone who votes must register first.

One's first impulse, in formalizing (1), is to treat it as a conditional statement, with antecedent "Something is broken" and consequent "Someone will pay for it."

$$(1') \quad (\exists x)(Bx) \rightarrow (\exists y)(Pyx)$$

After all, a slightly simpler statement, "If something is broken, someone will be fined" would be "similarly" symbolized:

$$(\exists x)(Bx) \rightarrow (\exists y)(Fy)$$

But although this last formula works, (1') does not. (1') fails because it does not convey the intended cross reference. The 'it' in the consequent of (1) refers back to the "something" of its antecedent, but the 'x' in the consequent of (1') "dangles": it cannot refer back to the 'x' in its antecedent because that antecedent is closed; the scope of the quantifier '$(\exists x)$' is just 'Bx'.

To convey the sense of (1) we must stretch the scope of the "x" quantifier in (1') so as to "capture" the dangling 'x' of the consequent:

$$(1'') \quad (\forall x)[Bx \rightarrow (\exists y)(Pyx)]$$

Similarly, of course, the sentence about fines could be rewritten:

$$(\forall x)[Bx \rightarrow (\exists y)(Fy)]$$

but there is no need; either version will do. With respect to (1), the rewriting is needed. Whereas (1') is ungrammatical – an open sentence with free variable 'x', masquerading as a sentence, and interchangeable with, say,

$$(\exists z)(Bz) \rightarrow (\exists y)(Pyx)$$

where there is no pretense of cross reference – (1″) is grammatical (a sentence rather than an open sentence) and conveys the sense of (1). We can think of the move from (1′) to (1″) as "justified" by Rule 7 of the rules of passage:

(7) $[(\exists x)(Fx) \to p] \leftrightarrow (\forall x)(Fx \to p)$

in analogy with the move from our first version of the sentence about fines to the second. But it is odd to think of the relation between the open, ungrammatical (1′) and the grammatical sentence (1″) as a relation of *equivalence*. (1″) actually conveys what (1′) only tries to convey. It is more reasonable to take a hard look at the original English sentence (1) and recognize its inherent universality. (1) tells us about any old thing (x) that, if it is broken, someone will pay for it; so it is appropriate to use an outside universal quantifier. We should remind ourselves that "ordinary" universals are often framed in existential language. "If someone is a citizen, he or she can vote" and "All citizens can vote" are interchangeable in English.

Note further that the contrasting existential "formulation" for (1):

(1‴) $(\exists x)[Bx \to (\exists y)(Pyx)]$

would be quite wrong. (1‴) tells us that there is something such that if it is broken, someone will pay for it, or, perhaps, that if certain things are broken, someone will pay for them. One can imagine a situation in which some special Chinese vase will be paid for if it is broken, but other breakage might be disregarded. Neither (1′) nor (1‴) correctly formalizes (1); only (1″) works.

Formalization of (2) is handled similarly.

(2) Everyone who votes must register first

We start by considering

$(\forall x)(Vx \to Rx)$

Everyone who votes must register.

This is grammatical, but inadequate; for the comparison expressed in (2) by the word 'first' has been lost. (2) tells us that the registering must occur before the voting; so the definitions used above:

Let $Vx = x$ votes.
Let $Rx = x$ registers.

do not suffice. We will need a reference to time, for purposes of comparison.

Let $Vxy = x$ votes at time y.
Let $Rxy = x$ registers at time y.

and also

Let $Bxy = x$ is before y.

Next we paraphrase (2):

Anyone who votes at some time or other must register at some time before that.

or, perhaps,

Anyone who votes at any time must register at some time before that.

These two paraphrases clearly come to the same thing, but they suggest two different strategies for formalizing:

$(\forall x)[(\exists y)(Vxy) \to x$ registers at a time z which is before $y]$

and

$(\forall x)(\forall y)(Vxy \to x$ registers at a time z which is before $y)$

The first strategy makes the same mistake that was made in (1'): the 'y' of the consequent cannot refer back to the 'y' of the antecedent (the voting time) because the quantifier '$(\exists y)$' covers the antecedent only, and so the intended cross reference is lost. The second strategy corrects this mistake, replacing the narrow-scope existential quantifier '$(\exists y)$' with the wide-scope universal '$(\forall y)$'. So we opt for the second version, and we arrive at

(2') $(\forall x)(\forall y)[Vxy \to (\exists z)(Rxz \land Bzy)]$

Finally we can – if we wish – stretch the scope of the other (inside) quantifier in each of (1) and (2) and derive two formulas in *prenex* form, with all the quantifiers out to the left, *preceding* the *nexus*, or open sentence. We use the rule of passage 6 of chapter 4:

$$[p \to (\exists x)(Fx)] \leftrightarrow (\exists x)(p \to Fx)$$

and get

(1a) $(\forall x)(\exists y)(Bx \to Pyx)$
(2a) $(\forall x)(\forall y)(\exists z)[Vxy \to (Rxz \wedge Bzy)]$

It is worth observing that the rules of passage can be used to "prenex" any statement, but that their use in the opposite direction – in the direction of "pure" form, or narrow scope of the quantifiers – is limited by the requirements of cross reference, as we have just seen.

It is also worth observing that each move by means of the rules of passage must preserve the order of the nested quantifiers. The rules of passage are, after all, equivalences which warrant the stretching or shrinking of the scope of one quantifier at a time. Such formulations as

$$(\exists y)(\forall x)(Bx \to Pyx)$$

or

$$(\forall x)(\exists z)(\forall y)[Vxy \to (Rxz \wedge Bzy)]$$

where the inside existential quantifiers have been moved to the left, "skipping" over a universal quantifier, would have been wrong. As we have seen earlier, pairs of like quantifiers (both existential or both universal) adjacent to each other are reversible; pairs of unlike quantifiers ($\forall \exists$ or $\exists \forall$) are emphatically not. If any equivalences are used to justify adjustment in the order of nested quantifiers, they must be used explicitly.

Prenex form turns out to be convenient for many purposes. For one, it makes it possible to apply UI and EI routinely in deductions. Prenex form is, however, notoriously difficult to read; it is much easier to see what (1″) and (2′) mean by just looking at them than it is to read (1a) and (2a).

Order of Nested Quantifiers

We turn next to the problem of ∀∃ – ∃∀ discrimination. Like such sentences as "All men are not agreeable" – which, as we have seen (at the end of chapter 5) are hopelessly ambiguous – many ordinary-language sentences that involve nested quantifiers are likewise ambiguous; the reader must rely upon intuition or context to decide what they mean. For example, Abraham Lincoln once said, "[Y]ou may fool all the people some of the time; you can even fool some of the people all the time; but you can't fool all of the people all the time."[4] Did Lincoln mean (in his first clause) that each of us can occasionally be fooled, or did he mean that on some occasions you can fool everybody? If he was thinking about what people are like, about the fact that all of us are fallible, he may have meant the first. If he was focusing on political situations – saying that some such situations are thoroughly manipulable – he may have meant the second. Again, if he was being cautious, he would have meant the first; for the first, being an ∀∃ statement, is weaker than the second. If he was making a bold statement (this is suggested, perhaps, by the word "even"), he would have meant the second. Let us formalize:

Let $Px = x$ is a person.
Let $Ox = x$ is an occasion.
Let $Fxy = x$ can be fooled at y.

The first clause can then be read:

A $(\forall x)[Px \to (\exists y)(Oy \wedge Fxy)]$
Every person can be fooled on some occasion or other.

or

A' $(\exists y)[Oy \wedge (\forall x)(Px \to Fxy)]$
There are occasions when everyone can be fooled.

---

[4] Quoted in Bartlett: *Familiar Quotations*, p. 542; spoken to a caller at the White House.

Similarly, the second clause can be read:

B $(\forall y)[Oy \rightarrow (\exists x)(Px \land Fxy)]$
On all occasions, someone or other can be fooled.

or

B′ $(\exists x)[Px \land (\forall y)(Oy \rightarrow Fxy)]$
There are some people who can be fooled on all occasions.

If Lincoln was focusing on people, he would have meant A and B′; if he was focusing on political situations he would have meant A′ and B. If he was being cautious he would have meant A and B; if he was being bold he would have meant A′ and B′. There are four possibilities. How do we decide what Lincoln meant?

The third clause presents no such ambiguity; it can be written in any number of ways, but they are all equivalent. The reader may wish to convince herself of this and also to check on whether the third clause conflicts with any version of the other two. I want instead to return to A and A′ and demonstrate my hitherto unsupported claim that A′ implies A.

A′ $(\exists y)[Oy \land (\forall x)(Px \rightarrow Fxy)]$

To prove:

A $(\forall x)[Px \rightarrow (\exists y)(Oy \land Fxy)]$

On strategy: to use a quantified statement, start by instantiating, and perhaps simplify, so as to see what you have to work with; to prove a quantified statement, choose an instance and plan to prove that, with an eye to generalizing at the end.

```
* 1  (∃y)[Oy ∧ (∀x)(Px → Fxy)]      PREM
* 2  Oa ∧ (∀x)(Px → Fxa)            1 EI (a)
* 3  Oa                             2 SIMP
* 4  (∀x)(Px → Fxa)                 2 SIMP
```

Now, planning our deduction from the bottom up, and aiming to prove an instance of the conclusion: that is, aiming to prove $Pb \rightarrow (\exists y)(Oy \land Fby)$ – not: $Pa \rightarrow (\exists y)(Oy \land Fay)$, because we are going to use UG in generalizing at the end and a has already been flagged at step 2 – we take its antecedent *Pb* as a premise, for conditional proof:

**	5 *Pb*	PREM
**	6 *Pb* → *Fba*	4 U1
**	7 *Fba*	6, 5 MP
**	8 *Oa* ∧ *Fba*	3, 7 ADJ
**	9 (∃*y*)(*Oy* ∧ *Fby*)	8 EG
*	10 *Pb* → (∃*y*)(*Oy* ∧ *Fby*)	5–9 CP

And we finish the deduction as planned:

*	11 (∀*x*)[*Px* → (∃*y*)(*Oy* ∧ *Fxy*)]	10 UG (*b*)
		QED

Finally, noting that there are two flagged steps in the deduction, steps 2 and 11, we check for double flagging and cross flagging. Neither *a* nor *b* is flagged twice; at step 2, where *a* is flagged, *b* does not occur; at step 11, where *b* is flagged, *a* does not occur; the deduction goes through.

What about the converse? Intuitively – and remembering the fallacious ∀∃ → ∃∀ argument about love – we observe that it seems quite possible for everyone to be gullible on occasion without there ever being a situation in which everyone is fooled at once. So the converse (A ∴ A′) seems to be invalid. Similarly for B ∴ B′.

Some statements involving nested quantifiers – like Lincoln's – are hopelessly ambiguous, unless one has definite information (or a definite theory) about the attitudes and interests of the speaker, which would serve to make his meaning clear. Others, for example,

There is a student who is bored in every one of his classes.

(Let *Bxy* = *x* is bored in *y*.)

(∃*x*)[*Sx* ∧ (∀*y*)(*Cy* → *Bxy*)]

are unambiguous. But one must be careful. The following is also unambiguous, and sounds in English very much like the statement about the student:

There is a post office in every city.

But it would be silly to formalize this sentence according to the pattern shown above (letting $Lxy = x$ is located in y):

$$(\exists x)[Px \land (\forall y)(Cy \to Lxy)]$$

That formula tells us that there is a post office that is located in all cities – it is in Chicago *and* in Los Angeles, for example – whereas what the English sentence tells us is that every city has a post office located in it, or

$$(\forall y)[Cy \to (\exists x)(Px \land Lxy)]$$

In other words, in reading such sentences as these, with an eye to formalizing them for logical purposes (even those which seem to us perfectly clear), we must look not only to linguistic clues – word order, phrases like "or other," and so forth – but also to common sense, shared social information, and context.

Reminders for Chapter 6

The Restrictions on EI and UG

The two restricted rules of inference for quantificational logic, Existential Instantiation (EI) and Universal Generalization (UG), require that the pseudonames introduced by EI or eliminated by UG be *flagged* at the step at which the rule is used and that the flagged letters be used with care.

A "real name" may not be flagged. A real name is a noun or letter assigned to an entity in the "real world" and used with that understanding in the premises or conclusion of an argument.

A pseudoname introduced by EI must be *new*, and a pseudoname eliminated by UG may *not* refer to a *special* case. These requirements are spelled out in the following restrictions:

1. A deduction in which any flag occurs must be *finished*; that is, no flagged letter may occur either in its conclusion or in any of its undischarged premises.
2. Within a deduction, no letter may be flagged twice. Using the same flag twice is the *fallacy of double flagging*.
3. Within a deduction, if a pseudoname occurs at a step at which another pseudoname is flagged, the first pseudoname may not be flagged at any step at which the other pseudoname occurs. Violation of this restriction is the *fallacy of cross flagging*.
4. Analogously, *cyclical flagging* is also prohibited. If a occurs where b is flagged, and b occurs where c is flagged, out to some pseudoname n, n may not occur where a is flagged. Violation of this restriction is the *fallacy of cyclical flagging*.

Q-Rules of Inference

Universal Instantiation: From a universally quantified statement any of its instances, without restriction, follows.

Existential Generalization: An existentially quantified statement is deducible from any of its instances, without restriction.

Existential Instantiation: From an existentially quantified statement we may derive an instance, provided that the pseudoname that replaces the variable of quantification is *new*.

Universal Generalization: From a statement containing a pseudoname we may derive a universally quantified statement of which it is an instance, provided that the pseudoname, which is replaced by the variable of quantification, does *not* refer to a *special* case.

Reminders

Q-INTerchange: Any two open sentences whose instances have been shown to be truth-functionally equivalent are interchangeable in quantificational contexts.

Change of Variable (CV): Any two quantified statements alike except for using different variables of quantification are equivalent and, accordingly, interchangeable.

Quantifier Negation

QN: $\sim(\forall x)(Fx) \leftrightarrow (\exists x)(\sim Fx)$
$\sim(\exists x)(Fx) \leftrightarrow (\forall x)(\sim Fx)$

QN': $\sim(\forall x)(Fx \rightarrow Gx) \leftrightarrow (\exists x)(Fx \wedge \sim Gx)$
$\sim(\exists x)(Fx \wedge Gx) \leftrightarrow (\forall x)(Fx \rightarrow \sim Gx)$

Other Equivalences

$[(\forall x)(Fx) \wedge (\forall x)(Gx)] \leftrightarrow (\forall x)(Fx \wedge Gx)$
$[(\exists x)(Fx) \vee (\exists x)(Gx)] \leftrightarrow (\exists x)(Fx \vee Gx)$

$(\forall x)(\forall y)(Fxy) \leftrightarrow (\forall y)(\forall x)(Fxy)$
$(\exists x)(\exists y)(Fxy) \leftrightarrow (\exists y)(\exists x)(Fxy)$

Rules of Passage

Where p is any sentence that does *not* contain x,

(1) $[p \wedge (\forall x)(Fx)] \leftrightarrow (\forall x)(p \wedge Fx)$

(2) $[p \wedge (\exists x)(Fx)] \leftrightarrow (\exists x)(p \wedge Fx)$

(3) $[p \vee (\forall x)(Fx)] \leftrightarrow (\forall x)(p \vee Fx)$

(4) $[p \vee (\exists x)(Fx)] \leftrightarrow (\exists x)(p \vee Fx)$

(5) $[p \rightarrow (\forall x)(Fx)] \leftrightarrow (\forall x)(p \rightarrow Fx)$

(6) $[p \rightarrow (\exists x)(Fx)] \leftrightarrow (\exists x)(p \rightarrow Fx)$

(7) $[(\exists x)(Fx) \rightarrow p] \leftrightarrow (\forall x)(Fx \rightarrow p)$

(8) $[(\forall x)(Fx) \rightarrow p] \leftrightarrow (\exists x)(Fx \rightarrow p)$

Problems for Chapter 6

I. In the Lincoln quotation, we discerned two versions of his second clause: "you can even fool some of the people all the time":

$$B \ (\forall y)[Oy \rightarrow (\exists x)(Px \land Fxy)]$$
$$B' \ (\exists x)[Px \land (\forall y)(Oy \rightarrow Fxy)]$$

and we conjectured that B does not imply B'. Here is a "proof" that B does imply B'.

*	1	$(\forall y)[Oy \rightarrow (\exists x)(Px \land Fxy)]$	PREM
*	2	$Oa \rightarrow (\exists x)(Px \land Fxa)$	1 UI
**	3	Oa	PREM
**	4	$(\exists x)(Px \land Fxa)$	2,3 MP
**	5	$Pb \land Fba$	4 EI (b)
**	6	Pb	5 SIMP
**	7	Fba	5 SIMP
*	8	$Oa \rightarrow Fba$	3–7 CP
*	9	$(\forall y)(Oy \rightarrow Fby)$	8 UG (a)
*	10	$Pb \land (\forall y)(Oy \rightarrow Fby)$	6,9 ADJ
*	11	$(\exists x)[Px \land (\forall y)(Oy \rightarrow Fxy)]$	10 EG

In ordinary language:

*	1	On all occasions someone can be fooled.	PREM
*	2	If Election Day (a) is an occasion, someone can be fooled on Election Day.	1 UI
**	3	Let's suppose that Election Day is an occasion.	PREM
**	4	In that case, someone can be fooled on Election Day.	2, 3 MP
**	5	Let's call him Joe Blow (b); Joe Blow is a person who can be fooled on Election Day.	4 EI (b)
**	6	Joe Blow is a person.	5 SIMP
**	7	Joe Blow can be fooled on Election Day.	5 SIMP

* 8 If Election Day is an occasion (see step 3),
     Joe Blow can be fooled on Election Day
     (see step 7).   3–7 CP
* 9 Generalizing, Joe Blow can be fooled
     on all occasions.   8 UG (a)
* 10 Joe Blow is a person who can be fooled
     on all occasions.   6, 9 ADJ
* 11 There is someone who can always be fooled. 10 EG

There are *two* fallacies in this "deduction." Identify and explain them both.

II. Do deductions in support of the following nine arguments:

1. There is a Cretan who lies to everyone.
   Therefore, there is a Cretan who lies to himself.

2. Everyone who hurts everyone who hurts him, hurts himself.
   Therefore, someone hurts somebody.

3. All horses are animals.
   Therefore, all heads of horses are heads of animals.

4. Anyone who trusts everyone is a fool.
   Anyone who trusts no one is a cynic.
   Not everyone is either a fool or a cynic.
   Therefore, some (people) trust some (people) but distrust others.

   (Note: the word "others" is to be disregarded here, as a "rhetorical flourish"; 'other' is not yet part of our logical vocabulary.)

5. All the world loves a lover.
   Mary loves someone who does not love her.
   Therefore, I will eat my hat.

6. There is a stove in every kitchen.
   Therefore, if there are no stoves, nobody eats in the kitchen.
7. There is a movie that every New Yorker wants to see.
   Therefore, every New Yorker wants to see a movie.
8. Joe approves of anyone who is demanding of himself.
   Joe disapproves of anyone who is demanding of everyone who works for her.
   Therefore, if anyone is demanding of everyone who works for her, someone does not work for herself.
9. Joe approves of anyone who is demanding of himself.
   Joe disapproves of anyone who is demanding of everyone who works for her.
   Therefore, if anyone is demanding of everyone who works for her, she does not work for herself.

Chapter 7
Truth Trees for Relational Logic

As the reader has surely become aware in dealing with the problems of chapter 6, the first difficulty encountered by the student of relational logic is that of restating ordinary-language sentences faithfully in symbolic form. Nesting of quantifiers introduces additional complexities in the building of truth trees. Where possible, it will be helpful to do some of the necessary instantiation before beginning to make a diagram. We deal first with a few examples.

It has already been shown, by deduction, that argument (2) of chapter 6 is valid; we now confirm this result by truth tree.

All circles are plane figures.
Therefore, everyone who draws a circle draws a plane figure.

PREM: $(\forall x)(Cx \to Px)$
CON: $(\forall x)[(\exists y)(Cy \land Dxy) \to (\exists z)(Pz \land Dxz)]$
~CON: $\sim(\forall x)[(\exists y)(Cy \land Dxy) \to (\exists z)(Pz \land Dxz)]$
$(\exists x)[(\exists y)(Cy \land Dxy) \land \sim(\exists z)(Pz \land Dxz)]$ by QN'
$(\exists x)[(\exists y)(Cy \land Dxy) \land (\forall z)(Pz \to \sim Dxz)]$ by QN'
$(\exists y)(Cy \land Day) \land (\forall z)(Pz \to \sim Daz)$ by ET (a)

So, simplifying, we see that the tree will contain representatives of the following:

~CON A: $(\exists y)(Cy \land Day)$ (Albert draws a circle)
~CON B: $(\forall z)(Pz \to \sim Daz)$ (No plane figures are drawn by Albert)
PREM: $(\forall x)(Cx \to Px)$

We begin with the existential, and then use the universals in whatever order:

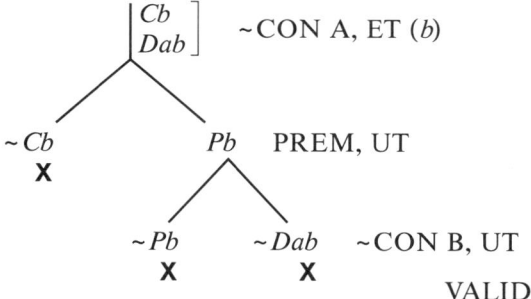

VALID

Two other instances: $Ca \to Pa$, from the premise by UT, and $Pa \to {\sim}Daa$, from the denial of the conclusion, were available, but not needed (in this case those extra instances wouldn't have made much sense), because the truth tree closed without them.

We turn next to another example from chapter 6:

Everyone who votes must register first.
Therefore, everyone who votes must register.

This looks like a valid argument, and we have already seen, separately, how both premise and conclusion may be formalized:

$$(\forall x)(\forall y)[Vxy \to (\exists z)(Rxz \land Bzy)]$$
$$\therefore (\forall x)(Vx \to Rx)$$

But, not surprisingly, this version of the conclusion will not be appropriate here, for it does not gibe with the premise. So we try again:

$$(\forall x)(\forall y)[Vxy \to (\exists z)(Rxz \land Bzy)]$$
$$\therefore (\forall x)(\forall y)(Vxy \to Rxy)$$

This is worse, for it does not convey what was intended by the conclusion. What it says is that everyone who votes at any time must register at that time: if Joe votes on Election Day, he must register on Election Day. This does not seem to be what was intended – and it surely would not follow from this premise. So, with a little more care, we paraphrase the conclusion:

Everyone who votes at some time or other must register at some time or other.

$$(\forall x)[(\exists y)(Vxy) \to (\exists z)(Rxz)]$$

Truth Trees for Relational Logic 259

PREM: $(\forall x)(\forall y)[Vxy \to (\exists z)(Rxz \land Bzy)]$
~CON: $\sim(\forall x)[(\exists y)(Vxy) \to (\exists z)(Rxz)]$
  $(\exists x) \sim [(\exists y)(Vxy) \to (\exists z)(Rxz)]$ by QN
  $(\exists x)[(\exists y)(Vxy) \land \sim(\exists z)(Rxz)]$ by QN'
  $(\exists x)[(\exists y)(Vxy) \land (\forall z)(\sim Rxz)]$ by QN
  $(\exists y)(Vay) \land (\forall z)(\sim Raz)$ by ET (a)
PREM': $(\forall y)[Vay \to (\exists z)(Raz \land Bzy)]$ by UT (using a)

The tree will contain representatives of the following:

~CON A: $(\exists y)(Vay)$ (There is a time when Albert votes)
~CON B: $(\forall y)(\sim Ray)$ (There is no time when Albert registers)
PREM': $(\forall y)[Vay \to (\exists z)(Raz \land Bzy)]$

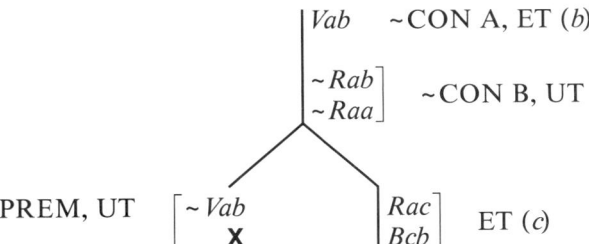

This looks like an open tree, signifying the invalidity of the argument – but it is not, for the universal ~CON B has not been fully utilized. The completed tree will look like this:

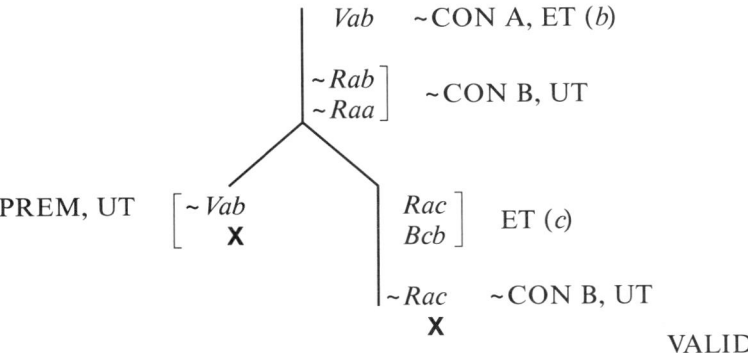

VALID

In this example, we could guard against such omission by entering the premise, which contains a nested existential, before the universal ~CON B:

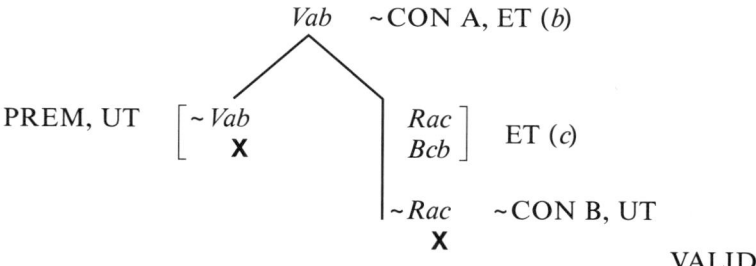

PREM, UT $\begin{bmatrix} \sim Vab \\ \mathbf{X} \end{bmatrix}$ $\quad\quad \begin{bmatrix} Rac \\ Bcb \end{bmatrix}$ ET (c)

$\qquad\qquad\qquad\qquad\quad \begin{vmatrix} \sim Rac \\ \mathbf{X} \end{vmatrix}$ ~CON B, UT

$\qquad\qquad\qquad\qquad\qquad\qquad\qquad\qquad$ VALID

This is good policy, but it cannot always be followed. Also, as we have seen, neglect of this policy need not be misleading, for we can always return to a universal, even if it has been entered prematurely, and use it again, as we did with the first tree.

We turn now to two related invalid arguments: the misinterpretation of this argument, and the converse.

First, the misinterpretation:

 Everyone who votes at some time must register first.
 Therefore, everyone who votes at some time must register then.

 PREM: $(\forall x)(\forall y)[Vxy \rightarrow (\exists z)(Rxz \land Bzy)]$
 CON: $(\forall x)(\forall y)(Vxy \rightarrow Rxy)$
 ~CON: $\sim(\forall x)(\forall y)(Vxy \rightarrow Rxy)$
    $(\exists x) \sim (\forall y)(Vxy \rightarrow Rxy)$ by QN
    $(\exists x)(\exists y)(Vxy \land \sim Rxy)$ by QN'
    $(\exists y)(Vay \land \sim Ray)$ by ET (a)
    $Vab \land \sim Rab$ by ET (b)
 PREM: $(\forall y)[Vay \rightarrow (\exists z)(Raz \land Bzy)]$ by UT
    $Vab \rightarrow (\exists z)(Raz \land Bzb)$ by UT

Truth Trees for Relational Logic

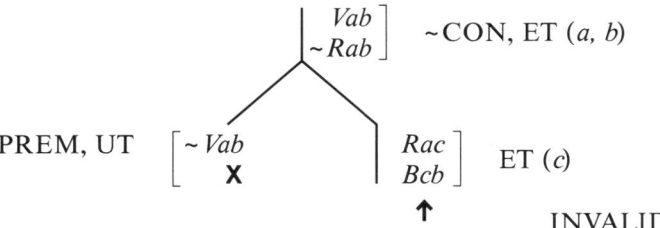

Given the premise, it is possible that Alice will vote at time b and register at time c, before b, and will not register at time b. Rac and $\sim Rab$ do not contradict each other.

Now, the converse:

Everyone who votes must register.
Therefore, everyone who votes must register first.

PREM:	$(\forall x)[(\exists y)(Vxy) \to (\exists z)(Rxz)]$	
$\sim$CON:	$\sim(\forall x)(\forall y)[Vxy \to (\exists z)(Rxz \wedge Bzy)]$	
	$(\exists x) \sim (\forall y)[Vxy \to (\exists z)(Rxz \wedge Bzy)]$	by QN
	$(\exists x)(\exists y)[Vxy \wedge \sim(\exists z)(Rxz \wedge Bzy)]$	by QN'
	$(\exists x)(\exists y)[Vxy \wedge (\forall z)(Rxz \to \sim Bzy)]$	by QN'

Instantiating,

	$(\exists y)[Vay \wedge (\forall z)(Raz \to \sim Bzy)]$	by ET (a)
	$Vab \wedge (\forall z)(Raz \to \sim Bzb)$	by ET (b)
PREM':	$(\exists y)(Vay) \to (\exists z)(Raz)$	by UT
	$(\forall y)(\sim Vay) \vee (\exists z)(Raz)$	by IF, QN

The tree will contain:

$\sim$CON A: Vab
$\sim$CON B: $(\forall z)(Raz \to \sim Bzb)$
PREM': $(\forall y)(\sim Vay) \vee (\exists z)(Raz)$

262 Relational Logic

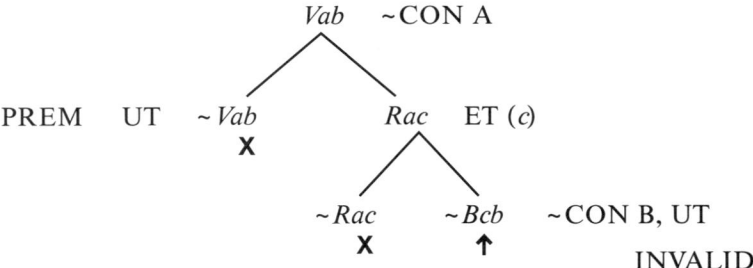

Given the premise, it is possible that Alice will vote at time b and register at time c, before b; she might register afterwards, or she might register when she votes. It should be remembered that the fact that the voting time b and the registering time c are denoted by different letters is no guarantee that they are different times: maybe so, maybe not. Also, nothing in either of these sentences guarantees that Alice does not register more than once, or vote more than once.

Finally, for completeness, we do a deduction for the original argument:

$$(\forall x)(\forall y)[Vxy \rightarrow (\exists z)(Rxz \land Bzy)]$$
$$\therefore (\forall x)[(\exists y)(Vxy) \rightarrow (\exists z)(Rxz)]$$

We plan to show: $(\exists y)(Vay) \rightarrow (\exists z)(Raz)$

*	1	$(\forall x)(\forall y)[Vxy \rightarrow (\exists z)(Rxz \land Bzy)]$	PREM
**	2	$(\exists y)(Vay)$	PREM (for CP)
**	3	Vab	2 EI (b) [a]
**	4	$(\forall y)[Vay \rightarrow (\exists z)(Raz \land Bzy)]$	1 UI
**	5	$Vab \rightarrow (\exists z)(Raz \land Bzb)$	4 UI
**	6	$(\exists z)(Raz \land Bzb)$	5, 3 MP
**	7	$Rac \land Bcb$	6 EI (c) [a,b]
**	8	Rac	7 SIMP
**	9	$(\exists z)(Raz)$	8 EG
*	10	$(\exists y)(Vay) \rightarrow (\exists z)(Raz)$	2–9 CP
*	11	$(\forall x)[(\exists y)(Vxy) \rightarrow (\exists z)(Rxz)]$	10 UG (a)
			QED

Next, we check for flagging. At step 3, where *b* is flagged, *a* occurs; at step 7, where *c* is flagged, both *a* and *b* occur; but at step 11, where *a* is flagged, neither *b* not *c* appears. No double flagging; no cross flagging; no cyclical flagging. The deduction is OK.

Infinite Trees

Let us look next at argument (3) of chapter 6:

$$(\exists x)(\forall y)(Lyx)$$
$$\therefore (\forall y)(\exists x)(Lyx)$$

PREM:	$(\exists x)(\forall y)(Lyx)$	
~CON:	$(\exists y) \sim(\exists x)(Lyx)$	by QN
	$(\exists y)(\forall x)(\sim Lyx)$	by QN
PREM':	$(\forall y)(Lya)$	from PREM, by ET (*a*)
~CON':	$(\forall x)(\sim Lbx)$	from ~CON, by ET (*b*)

$$\left.\begin{array}{l} Laa \\ Lba \end{array}\right] \quad \text{from PREM'} \\ \text{by UT}$$

$$\left.\begin{array}{l} \sim Lba \\ \sim Lbb \end{array}\right] \quad \text{from ~CON'} \\ \text{by UT}$$

X

We turn next to its converse, which we have already judged to be invalid.

$$(\forall y)(\exists x)(Lyx)$$
$$\therefore (\exists x)(\forall y)(Lyx)$$

PREM:	$(\forall y)(\exists x)(Lyx)$	
~CON:	$\sim(\exists x)(\forall y)(Lyx)$	
	$(\forall x) \sim(\forall y)(Lyx)$	by QN
	$(\forall x)(\exists y)(\sim Lyx)$	by QN

Let us look at these two formulas separately, each of them $\forall\exists$ in structure.

PREM: $(\forall y)(\exists x)(Lyx)$
$(\exists x)(Lax)$ by UT
Lab by ET (b)
$(\exists x)(Lbx)$ by UT
Lbc by ET (c)
$(\exists x)(Lcx)$ by UT
Lcd by ET (d)

The list will go on forever.

Similarly, for the denial of the conclusion:

~CON: $(\forall x)(\exists y)(\sim Lyx)$
$(\exists y)(\sim Lyb)$ by UT (trying to connect with the premise)
$\sim Lb'b$ by ET (b')
$(\exists y)(\sim Lyb')$ by UT
$\sim Lb''b'$ by ET (b'')
$(\exists y)(\sim Lyb'')$ by UT
$\sim Lb'''b''$ by ET (b''')

and so forth. This list: $\sim Lb'b$, $\sim Lb''b'$, $\sim Lb'''b''$, ..., like Lab, Lbc, Lcd, ..., will go on forever, and so will $Lb'c'$, $Lc'd'$, ... $Ld'e'$, generated by the premise, starting with b', and $\sim Lc''c'$, $\sim Lc'''c''$, $\sim Lc''''c'''$, ..., generated by the denial of the conclusion, starting with c'. The tree, which will consist of an endless list of statements of the forms Lyx and $\sim Lyx$, in conjunction, shows no sign of closing, even though, since some of these statements are affirmative and some are negative, it is possible that there is contradiction which is not apparent. On the other hand, the endlessness of this tree is no guarantee that $(\forall y)(\exists x)(Lyx)$ and $(\forall x)(\exists y)(\sim Lyx)$ are consistent, as we are, of course, inclined to believe.

So, instead of allowing our instances to proliferate, we imagine that the world consists of just two characters, a and b. If it is possible, in that little world, for $(\forall y)(\exists x)(Lyx)$ and $(\forall x)(\exists y)(\sim Lyx)$ to be true together, then neither one of them implies the denial of the other. Given just a and b, $(\forall y)(\exists x)(Lyx)$ reduces to the conjunction: $(\exists x)(Lax) \land (\exists x)(Lbx)$; and $(\exists x)(Lax)$ and $(\exists x)(Lbx)$ reduce, respectively, to the disjunctions: $Laa \lor Lab$ and $Lba \lor Lbb$. So the premise is

$$(Laa \lor Lab) \land (Lba \lor Lbb)$$

"Everyone loves someone or other" tells us that a loves a or b and that b loves a or b. Similarly, the denial of the conclusion ("Someone is loved by everybody"), that is, "Everyone is unloved by somebody or other," or $(\forall x)(\exists y)(\sim Lyx)$, reduces to $(\exists y)(\sim Lya) \land (\exists y)(\sim Lyb)$, and then to

$$(\sim Laa \lor \sim Lba) \land (\sim Lab \lor \sim Lbb)$$

which tells us that a is either unloved by a or unloved by b and that b is either unloved by a or unloved by b.

PREM: $(Laa \lor Lab) \land (Lba \lor Lbb)$
$\sim$CON: $(\sim Laa \lor \sim Lba) \land (\sim Lab \lor \sim Lbb)$

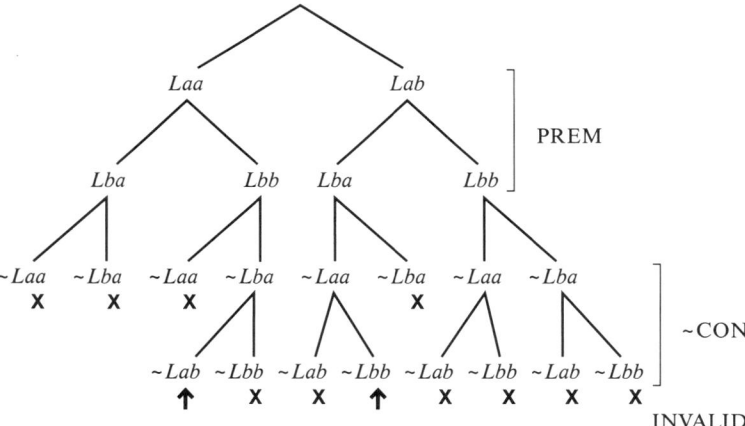

There are two open branches: Laa, Lbb, $\sim Lba$, $\sim Lab$ (a and b love themselves but not each other); and Lab, Lba, $\sim Laa$, $\sim Lbb$ (a and b love each other but not themselves).

This tree establishes that the argument we have been discussing is invalid; for it exhibits a counterexample to that argument, a situation in which the premise is true, but the conclusion is false.

But if the tree had closed, it would *not* have demonstrated that the argument was valid. After all, in a still smaller world, in which a was alone, the premise would have yielded Laa and the denial of the conclusion $\sim Laa$. That truth tree would close:

| *Laa* PREM
| ~*Laa* ~CON
| **X**

Narrowing the universe may help us to find counterexamples, but not to demonstrate validity.

The existence of infinite trees marks an important difference between relational logic and what has gone before. In propositional logic both truth tables and truth trees have yielded routine – though often cumbersome – methods for getting a yes-or-no answer to the question: Is this argument (or this proposed theorem) valid? Truth trees also yield such yes-or-no answers in predicate logic. But when we come to relations truth trees are no longer adequate in that way. A new level of complexity has been introduced.

If an argument is valid, its truth tree will close, although, even in such cases, attention to strategy is worth while and preferable to reliance on routine procedures. But, if an argument is invalid, the truth-tree method can no longer be counted on to establish that invalidity, by producing a counterexample that can be read off from a completed tree. Some truth trees for relational arguments cannot be completed.

An uncompleted tree will, of course, still be informative. It may, for example, suggest how an intelligent person could construct a counterexample to disprove an argument (or statement) that had been judged to be invalid. But it will not construct that counterexample for him.

What Can Be Gleaned from Truth Trees

Before going on to study the logic of identity, let us look at two more arguments from chapter 6: Problem I, concerning the quotation from Lincoln, and the argument from "All the world loves a lover."

The truth tree for the invalid argument of Problem I:

Truth Trees for Relational Logic 267

I. On all occasions someone or other can be fooled. Therefore, there are some people who can be fooled on all occasions.

$$(\forall y)[Oy \to (\exists x)(Px \land Fxy)]$$
$$\therefore (\exists x)[Px \land (\forall y)(Oy \to Fxy)]$$

like the truth tree for

$$(\forall y)(\exists x)(Lyx)$$
$$\therefore (\exists x)(\forall y)(Lyx)$$

will go on forever. Nonetheless, this more complicated tree can be used directly to show the invalidity of the argument.

PREM: $(\forall y)[Oy \to (\exists x)(Px \land Fxy)]$
~CON: $(\forall x)[Px \to (\exists y)(Oy \land \sim Fxy)]$

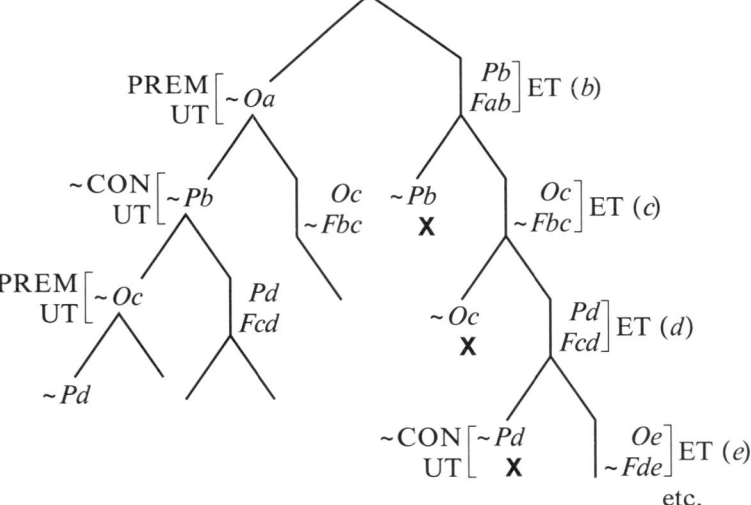

The middle of this tree proliferates unmanageably – at least from the point of view of a diagrammer. The right-hand branch yields a familiar-looking list that includes: *Fab*, ~*Fbc*, *Fcd*, ~*Fde*, ..., which is inconclusive, but which suggests a way of constructing a smaller tree that might yield a counterexample.

The far left-hand branch, however, is straightforwardly consistent and will remain consistent as it grows, since every item on that branch is negative. It tells us that if $(\forall y)(\sim Oy)$ and $(\forall x)(\sim Px)$, that is, if there were no occasions (political situations, if you will – it may be hard to imagine a world without times) and no people, the premise of this argument would be true and its conclusion false.

The student was asked in Problem I to find two fallacies in the deduction given:

*	1	$(\forall y)[Oy \to (\exists x)(Px \land Fxy)]$	PREM
*	2	$Oa \to (\exists x)(Px \land Fxa)$	1 UI
**	3	Oa	PREM
**	4	$(\exists x)(Px \land Fxa)$	2, 3 MP
**	5	$Pb \land Fba$	4 EI (b)
**	6	Pb	5 SIMP
**	7	Fba	5 SIMP
*	8	$Oa \to Fba$	3–7 CP
*	9	$(\forall y)(Oy \to Fby)$	8 UG (a)
*	10	$Pb \land (\forall y)(Oy \to Fby)$	6, 9 ADJ
*	11	$(\exists x)[Px \land (\forall y)(Oy \to Fxy)]$	10 EG

How does this tree suggest where those fallacies can be found?

The premise
$$(\forall y)[Oy \to (\exists x)(Px \land Fxy)]$$
will be true if there are no occasions, by virtue of the falsity of the antecedent of its open sentence, and it will still be true if there are no people, as the truth tree shows. But the conclusion
$$(\exists x)[Px \land (\forall y)(Oy \to Fxy)]$$
says that there are people, and so will be obviously false if there are none.

At what point in the deduction is a move made to import the claim that there are people? Where did the 'Px' of the conclusion come from?

In the "deduction" of Problem I, the conclusion is derived, correctly enough, from step 10: $Pb \land (\forall y)(Oy \to Fby)$

and the '*Pb*' of step 10 is derived in turn from step 6. But step 6, it turns out, is not available; its premise, '*Oa*' at step 3, was discharged at step 8. Step 10, in other words, commits the fallacy of illicit repetition.

A long story perhaps – but it goes to show that the far left branch of this truth tree not only establishes the invalidity of the argument, but also points to the fallacy of illicit repetition in the failed deduction. The premise gives us no guarantee that there are any occasions; if there are no occasions, it gives us no guarantee that there are any people; the conclusion says that there are people – so the argument fails.

The other fallacy, cross flagging at step 9, corresponds (very roughly) to the endlessness of the tree; this fallacy is easy for most students to discover, since cross flagging is discussed in chapter 6, whereas illicit repetition goes back to basic truth-functional principles from chapter 1, which may have been forgotten.

Now let us return to "All the world loves a lover":

All the world loves a lover.
Somebody loves somebody.
Therefore, everybody loves everybody.

PREM 1: $(\forall x)[(\exists y)(Lxy) \rightarrow (\forall z)(Lzx)]$
$(\forall x)[\sim(\exists y)(Lxy) \vee (\forall z)(Lzx)]$ by IF
$(\forall x)[(\forall y)(\sim Lxy) \vee (\forall z)(Lzx)]$ by QN
PREM 2: $(\exists x)(\exists y)(Lxy)$
~CON: $\sim(\forall x)(\forall y)(Lxy)$
$(\exists x) \sim(\forall y)(Lxy)$ by QN
$(\exists x)(\exists y)(\sim Lxy)$ by QN

We begin this tree by entering the two existentials: the second premise and the denial of the conclusion:

Lab | PREM 2, ET (*a*), (*b*)
$\sim Lcd$ | ~CON, ET (*c*), (*d*)

(a loves b, and c does not love d). In entering the first premise, using UT, we have some choice about how to begin. It will be sensible to try to contradict either Lab or $\sim Lcd$. Aiming first to contradict Lab, we look to the "antecedent" side, where the "not" signs are, and so we put a for x and b for y: If a loves b then everyone (here, a, b, c, and d) loves a. This closes the left-hand branch, but leaves the right-hand branch open:

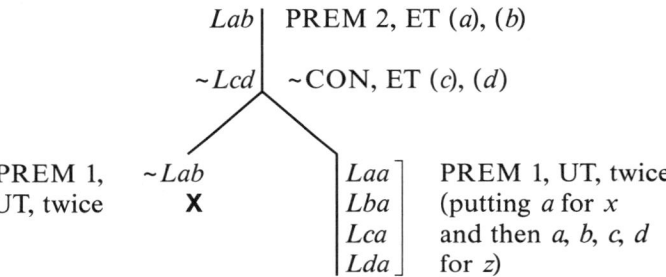

We have contradicted Lab; we want now to contradict $\sim Lcd$. So we look to the consequent side of the first premise, where the affirmatives are, and we put c for z and d for x. This closes the right-hand branch, leaving an "antecedent" branch: $\sim Lda$, $\sim Ldb$, $\sim Ldc$, $\sim Ldd$. We choose $\sim Lda$, and the tree closes:

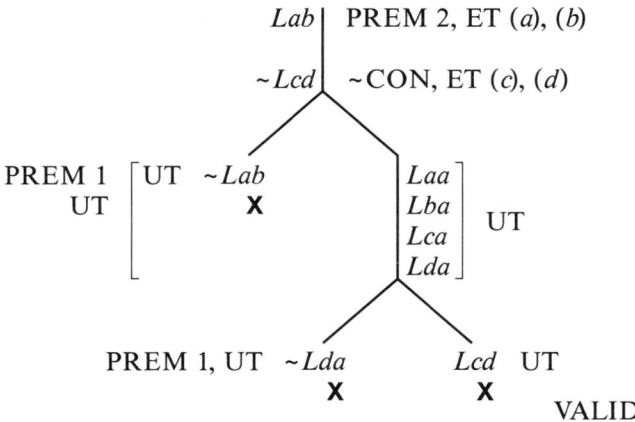

Truth Trees for Relational Logic   271

The "full" version of the first use of PREM 1, with *a* for *x*, is:

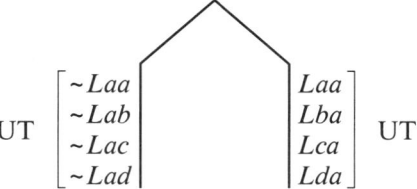

The "full" version of the second use, with *d* for *x*, is:

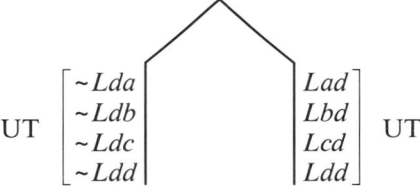

An unnecessarily lengthy but perhaps more perspicuous truth tree for this argument, therefore, would be like this:

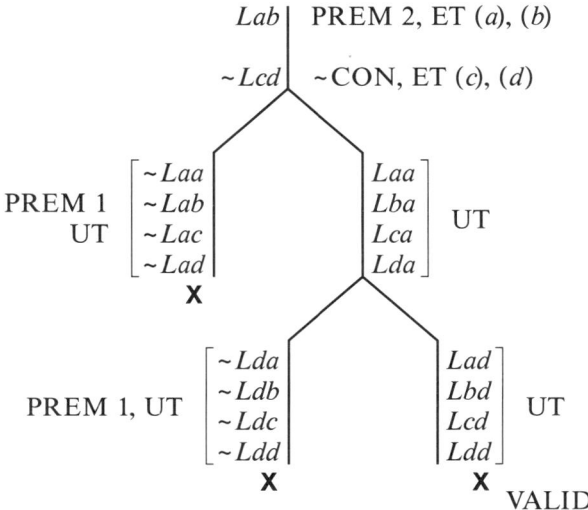

And there are two more needless uses. But there is no *one* use of Premise 1 that would be sufficient.

If you had trouble designing a deduction, because you were tempted to double-flag and did not know how to avoid it, the truth tree tells you how: use the first premise twice.

Reminders for Chapter 7

Basic truth trees:

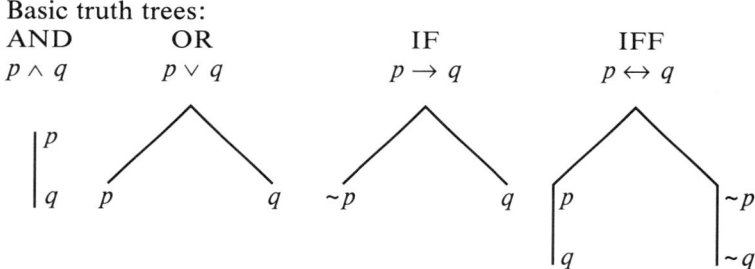

ET: In place of an existentially quantified sentence or component of a sentence to be entered on an open branch of a truth tree, enter an instance of it; in every case enter a *new* name.

UT: In place of a universally quantified sentence or component of a sentence to be entered on an open branch of a truth tree, enter every instance of it that uses a name or pseudoname already appearing on the branch in question.

See also Reminders for Chapter 6.

Problems for Chapter 7

I. Do truth trees for the arguments of Problem II of chapter 6.

II. Consider the following arguments. First, decide whether each argument is valid. Then confirm (or disconfirm) your judgment by truth tree, and explain the result. Do a deduction if valid.

1. There is a kitchen in every apartment.
   There are some apartments in which there is no comfortable place to eat.
   Therefore, some kitchens are not comfortable places to eat.

2. Everyone who hurts everyone who hurts him, hurts himself.
   Therefore, everyone hurts someone or other.

3. Everyone who hurts everyone who hurts her, hurts herself.
   Therefore, everyone is hurt by someone or other.

4. Anyone who trusts everyone is a fool.
   Anyone who trusts no one is a cynic.
   Not everyone is either a fool or a cynic.
   Therefore, some people are trusted by some and distrusted by others.

5. There is someone who will win by a wide margin if everyone who now supports her votes for her.
   Some people will not vote at all and no one will win by a wide margin.
   Therefore, some people will change their vote (that is, vote for someone they do not now support or fail to vote for someone they are now supporting).

6. Everyone who likes all Wagner, dislikes all Gershwin.
   Everyone who likes all Gershwin, dislikes all Wagner.
   Therefore, there is no one who likes some Wagner and some Gershwin.

III. With the understanding that to be a parent is to be parent to someone, that to have a parent is for someone to be parent to you, and that one's grandparent is parent to someone who is one's parent, consider the following twelve statements. With respect to each of these decide whether it is valid – a theorem of relational logic. If it is valid, do a deduction to establish it. If it is not valid, explain why not.

Begin by formalizing the definitions given above. In thinking about the validity of these statements be sure to set aside whatever else you know about parents and grandparents. These definitions do not, for example, immediately rule out the possibility that one could be one's own parent.

1. Everyone who is a parent has a parent.

2. If everyone is a parent, everyone has a parent.

3. If no one is a parent, no one has a parent.

4. If no one has a parent, no one is a parent.

5. Everyone who is a grandparent is a parent.

6. Everyone who has a parent has a grandparent.

7. If everyone has a grandparent, everyone has a parent.

8. If everyone has a parent, everyone has a grandparent.

9. If no one has a parent, no one has a grandparent.

10. If no one is his own parent, no one is his own grandparent.

11. If no one is her own grandparent, no one is her own parent.

12. If some men are not women, not everyone who has a father (a parent who is a man) has a mother (a parent who is a woman).

# Part IV
Identity and Description

Chapter 8
The Logic of Identity

> Were I the Moor, I would not be Iago.
> In following him, I follow but myself;
> ... I am not what I am.
> *Othello*, Act I, Scene 1.

The concept of identity is central to our ordinary ways of thinking, but it is more likely to be used implicitly than explicitly, and, more often than not, in the negative rather than the affirmative. We are more likely to inform someone that a is other than b than that a is identical with b. We use the concepts of identity and of otherness when we introduce our friends and relatives to one another, when we identify wrongdoers or exonerate the blameless, when we tell stories, when we write history, when we classify plants and animals, and when we diagnose diseases.

In chapter 6 of this book we have seen that such arguments as the following are valid:

> Mary is tolerant of everyone.
> Therefore, Mary is tolerant of herself.

$$\frac{(\forall x)(Tmx)}{\therefore Tmm}$$

That is, we know that if Mary is tolerant of everyone, she must be tolerant of herself. But if someone, say Henry, told us that Mary was tolerant of everyone but intolerant of herself, we would not charge *Henry* with inconsistency – even though we *would* charge Henry with inconsistency if he told us that Mary was tolerant of everyone but intolerant of *George*. In thinking about Mary's intolerance of herself we might regard *Mary* as inconsistent in her behavior or her attitudes, but we would be unlikely to regard Henry as inconsistent in his thinking. For we would hear him as saying, "Mary is tolerant of everyone *else*," which means that Mary is tolerant of everyone other than Mary. And *that* claim does not imply the conclusion that Mary is tolerant of herself (*Tmm*) and so does not conflict with Henry's "Mary is intolerant of herself" (*~Tmm*).

Linguistically, identity is troublesome in another way: it seems to seduce us into grammatical confusion. An English teacher explains, for example, that George Eliot was identical with Mary Ann Evans. "George Eliot and Mary Ann Evans were the same person," he says. But if "they were" the same person then there was only one of him (her?), and so what the teacher should have said was "George Eliot and Mary Ann Evans *was* the same person" – which sounds awful. I will try to avoid that sort of construction, but I do not expect to be able to do so. And I will occasionally use "scare quotes" to draw attention to idiomatic but misleading usage.

The concept of identity tends also to trap us in tautology or redundancy, both of which are often useful, but which are not useful when we are trying to convey information. There is a saying of Joseph Butler's, for example, which is often quoted by modern philosophers: "Everything is what it is, and not another thing."[1] The "and" in that quotation, like the "and" between 'George Eliot' and 'Mary Ann Evans', is a bit out of place, because the two clauses in the Butler quote are equivalent: they convey the same information, just as the two names denote the same British author. Otherness is the negation of identity; so the second clause is the double negation of the first.

To say of *a* and *b* that "they" are identical is to say that "they" are the same thing, that there is only one thing there – not that "they" are alike, or the same size, or the same shape, or the same value, or even indistinguishable. If *a* and *b* are indeed identical, it *follows* that "they" are alike, and the same size, and the same shape, and the same value, and indistinguishable. For "they" are not two, but one.

---

[1] Preface to *Fifteen Sermons Preached at the Rolls Chapel* (Oxford), Stanhope, 1792, par. 39. Variously reprinted; available in Joseph Butler, *Five Sermons* (New York: Liberal Arts Press, 1950), p. 15.

Principles of Identity

This is the ground for what has come to be known as "Leibniz's law"[2] (LL), one of the two principles of identity we shall be using in this chapter and the next.

*In*appropriately stated, Leibniz's law tells us that if any two things are identical, then whatever is true of one is true of the other. But, of course, if there is identity, there are not two of them, nor is there any other. What is other is a name or pseudoname, not an entity referred to – a linguistic expression, not a message conveyed.

Leibniz's law: Given a statement of identity, its terms are interchangeable within any statement containing either of those terms.

Accordingly: $\begin{cases} x=y \\ Fx \end{cases}$
∴ Fy LL

Likewise, $\begin{cases} x=y \\ Fy \end{cases}$
∴ Fx LL

Here we are using free variables 'x' and 'y' as we used 'p' and 'q' in Part I and 'F' and 'G' in Parts II and III, in the symbolic examples of the use of the rules. 'a' and 'b' will be appropriate in actual deductions or truth trees.

We have, according to tradition, borrowed the notation of arithmetic; we will use the equals sign '=' for "is identical with," and also the inequality sign '≠' for "is not identical with", or "is other than." (It would be unhelpful, however, to look to arithmetic for the meanings of those signs.) We will use '=' and '≠' to

---

[2] It is generally agreed that this title is a misnomer. For, although Leibniz may well have believed that Leibniz's law is true, he certainly did not take credit for its discovery or regard it as a principle of importance in his philosophy. He was indeed deeply interested in the concept of identity, but the principles he cared about were the identity of indiscernibles and his "Salva veritate" principle, which is the converse of Leibniz's law. See, for example, Hide Ishiguro, *Leibniz's Philosophy of Logic and Language* (Cambridge: University Press, 2d ed., 1990), pp. 17, 42. I do not know when or how the title came into common usage.

connect names or pseudonames (or noun variables), not statements or predicates.

We will use Leibniz's law as a principle of inference. Given that George Eliot was identical with Mary Ann Evans and that George Eliot wrote *Middlemarch*, Leibniz's law allows us to infer that Mary Ann Evans wrote *Middlemarch*.

We can also use Leibniz's law to deduce its negatively stated version, the derived rule LL′:

> Leibniz's law′: Given a pair of statements, one affirmed, one denied, which are otherwise exactly alike except for the interchange of a pair of terms, it follows that the items denoted by those terms are not identical.

Accordingly:

$$\begin{cases} Fx \\ \sim Fy \end{cases} \over \therefore x \neq y$$ LL′

(Compare the two rules of detachment, modus ponens and modus tollens, of chapter 1.) LL′ is the principle that justifies the courtroom strategy of the alibi (literally translated from the Latin, "otherwise"). If the murderer was in Times Square at noon on Saturday (committing the murder) and Joe Blow was not in Times Square at noon on Saturday (he was in San Francisco), then the murderer was not Joe Blow.

Here is a deduction of a representative instance of LL′:

```
 * { 1  Fa                           PREM
   { 2  ~Fb

            To Prove:  a≠b

** 3  a=b                            PREM
** 4  Fb                             3, 1 LL
** 5  Fb ∧ ~Fb                       4, 2 ADJ
 * 6  a≠b                            3-5 RED
                                     QED
```

The second principle we shall need in dealing with the concept of identity cannot be treated as a principle of inference; this is the principle of self-identity, or "Butler's law," BL:

Butler's law: Everything is identical with itself.

That is,

$$(\forall x)(x=x) \quad \text{BL}$$

The truth of this claim seems obvious; it is hard to imagine any grounds for questioning it, for it is expressive of the meaning of the only concept it involves, that of identity. We shall treat it as if it were a theorem. More precisely, we shall treat it as an *axiom*: it will be used as we use a theorem, but it requires no proof. Any instance of the axiom of self-identity may be introduced as a step in proof at any point in any deduction.

Some Deductions

Let us begin by proving that '=', like '↔', '∧', and '∨', is commutative, i.e., that, for example, '$a=b$' is equivalent to '$b=a$'. The statement '$a=b$' tells us about a that it (a) is identical with b and about b that a is identical with it. Likewise, '$b=a$' tells us about b that it is identical with a and about a that b is identical with it. (These "four" claims seem to come to the same thing.) The axiomatic statement '$a=a$', which we shall be using in the deduction, can be thought of in *three* ways: it tells us about a, first, that it is identical with a, i.e., that it "$= a$"; second, that a is identical with it, i.e., that "$a=$" it; and, finally, that it is self-identical. Let us focus on the first of these ways (the second, of course, would do as well).

Let 'Fx' stand for '$x=a$'.
Then 'Fa' will be '$a=a$'.
and 'Fb' will be '$b=a$'.

The following, accordingly, will be an instance of Leibniz's law:

$a=b$
$\underline{a=a}$
∴ $b=a$

$a=b$
$\underline{Fa}$
∴ Fb

Similarly, using 'Gx' for '$x=b$', the following will be an instance of Leibniz's law:

$$\frac{\begin{array}{c}b=a\\b=b\end{array}}{\therefore\ a=b} \qquad \frac{\begin{array}{c}b=a\\Gb\end{array}}{\therefore\ Ga}$$

The deduction, then, will look like this:

To Prove: $a=b \leftrightarrow b=a$

```
*  1  a=b                    PREM
*  2  a=a                    BL
*  3  b=a                    1, 2 LL
   4  a=b → b=a              1-3 CP
*  5  b=a                    PREM
*  6  b=b                    BL
*  7  a=b                    5, 6 LL
   8  b=a → a=b              5-7 CP
   9  a=b ↔ b=a              4, 8 ADJ, DEF ↔
                             QED
```

This theorem justifies the ordinary practice (which seems to me to be right) of freely interchanging such pairs of identity statements. After all, there is no difference in meaning between '$a=b$' and '$b=a$', and so an intelligent human being may naturally disregard the difference in word order. In contrast, a computer may have to be carefully programmed and do a certain amount of work if it is to achieve a comparable level of indifference. The theorem confirms the view that, like 'George Eliot' and 'Mary Ann Evans', '$a=b$' and '$b=a$' are two expressions for the same reality.

The first common notion for which the concept of identity is needed (the second is the number 1) is *otherness*, the notion of being something else. This is the simple denial of identity. We noted earlier (in connection with Problem II.4 of chapter 6) that our logical notation in earlier chapters was incapable of conveying the conceptual difference between saying that we trust some people and also distrust some and saying that we trust some and distrust *others*. We commented then that, for the time being, we could afford to disregard the difference. We know intuitively that those whom we distrust must be *others*. It was not misleading to disregard the difference between the two statements because they are equivalent. Let us now prove that the second of those statements follows from the first.

Let 'Txy' abbreviate 'x trusts y'.
(Then $\sim Txy = x$ distrusts y.)

PREM: Some people trust some people and distrust some:
$(\exists x)[(\exists y)(Txy) \wedge (\exists y)(\sim Txy)]$

CON: Some people trust some people and distrust others:
$(\exists x)(\exists y)(\exists z)(Txy \wedge \sim Txz \wedge y \neq z)$

Note that, whereas the premise may be written either in relatively pure form, as above, or prenexed:

$(\exists x)(\exists y)(\exists z)(Txy \wedge \sim Txz)$

[*not* $(\exists x)(\exists y)(Txy \wedge \sim Txy)$, which is self-contradictory], the conclusion, since it contains '$y \neq z$', requires wide scope.

It is obvious that the converse is valid:

$$\frac{(\exists x)(\exists y)(\exists z)(Txy \wedge \sim Txz \wedge y \neq z)}{\therefore \ (\exists x)(\exists y)(\exists z)(Txy \wedge \sim Txz)}$$

The reader may wish to carry out that deduction for himself. (It requires seven steps, not just one.)

We shall deduce the following:

$$\frac{(\exists x)[(\exists y)(Txy) \wedge (\exists y)(\sim Txy)]}{\therefore \ (\exists x)(\exists y)(\exists z)(Txy \wedge \sim Txz \wedge y \neq z)}$$

*	1	$(\exists x)[(\exists y)(Txy) \wedge (\exists y)(\sim Txy)]$	PREM
*	2	$(\exists y)(Tay) \wedge (\exists y)(\sim Tay)$	2 EI (a)
*	3	$(\exists y)(Tay)$	2 SIMP
*	4	$(\exists y)(\sim Tay)$	2 SIMP
*	5	Tab	3 EI (b)
*	6	$\sim Tac$	4 EI (c)
*	7	$b \neq c$	5, 6 LL'
*	8	$Tab \wedge \sim Tac \wedge b \neq c$	5, 6, 7 ADJ

*	9 $(\exists z)(Tab \land \sim Taz \land b \neq z)$	8 EG
*	10 $(\exists y)(\exists z)(Tay \land \sim Taz \land y \neq z)$	9 EG
*	11 $(\exists x)(\exists y)(\exists z)(Txy \land \sim Txz \land y \neq z)$	10 EG
		QED

The Number 1

We turn now to the number 1, beginning with some formalization.

Henry is in the room:
Rh

Only Henry is in the room:
No one other than Henry is in the room:

$\sim(\exists x)(Rx \land x \neq h)$

or, by QN',

$(\forall x)(Rx \to x=h)$

Henry and only Henry is in the room:
$Rh \land (\forall x)(Rx \to x=h)$

At least one person is in the room:
$(\exists x)(Rx)$

More than one person is in the room:
$(\exists x)(\exists y)(Rx \land Ry \land x \neq y)$

No more than one person is in the room:

$\sim(\exists x)(\exists y)(Rx \land Ry \land x \neq y)$

or

$(\forall x)(\forall y)[(Rx \land Ry) \to x=y]$

Exactly one person is in the room:
$(\exists x)(Rx) \land (\forall x)(\forall y)[(Rx \land Ry) \to x=y]$

This last statement can be written as shown, as a conjunction of two of the statements above it, or with wide scope, as follows:

or
$$(\exists x)[Rx \wedge \sim(\exists y)(Ry \wedge y \neq x)]$$
$$(\exists x)[Rx \wedge (\forall y)(Ry \to y=x)]$$

There is an interesting consolidation of the third and last of these formulas. But first, some background.

The Greeks treated what we have been calling "singular" propositions – "Henry is in the room," "Socrates is wise" – not as a separate category, but as universals, in contrast with particulars. For their purposes, "Socrates is wise" meant no more (or less) than "All Socrateses are wise" or "Everyone who is Socrates is wise," which we would now write '$(\forall x)(x=s \to Wx)$', and this analysis became part of the tradition of the Aristotelian syllogism.

In contrast, a modern student would be more likely to think of "Socrates is wise" as a "particular", that is, as "Somebody named 'Socrates' is wise" or "Someone who is Socrates is wise" or '$(\exists x)(x=s \wedge Wx)$'. In chapter 4 we warned the reader not to confuse such a sentence as "Socrates is wise" – Ws – with the less informative existential "Someone is wise" or '$(\exists x)(Wx)$'. That warning was needed.

Interestingly enough, however, it turns out that, once the concept of identity is introduced, along with the notation needed to deal with it, both the Aristotelian and the mythical "modern student" are *right*.

"Socrates is wise" is indeed equivalent both to "Everyone who is Socrates is wise" and to "Someone who is Socrates is wise."[3] In other words, the following three equivalences hold:

$$Ws \leftrightarrow (\forall x)(x=s \to Wx)$$
$$Ws \leftrightarrow (\exists x)(x=s \wedge Wx)$$
$$(\forall x)(x=s \to Wx) \leftrightarrow (\exists x)(x=s \wedge Wx)$$

---

[3] For an insightful discussion of the relation between singular statements on the one hand and universal and particular statements on the other, in the traditional mode, see Fred Sommers, "Do We Need Identity?" *Journal of Philosophy* LXVI, 15 (Aug. 7, 1969): 499–504. Sommers makes the ingenious suggestion that a singular statement should be treated as "quantity-wild" (page 502), on the analogy of "wild" cards in poker. The equivalences here stated justify Sommers' strategy.

This is an odd result. The third equivalence is especially surprising, since it claims that a universal is equivalent to an existential – which is *not* a usual arrangement. It depends, of course, upon the special properties of the "relation" of identity.

Let us return now to the first equivalence, which is the one that will be useful in the consolidation of '$Rh \wedge (\forall x)(Rx \rightarrow h=x)$'. We shall do a deduction for that equivalence and leave the other two deductions to the reader (see Problems I.3 and I.4 below).

To Prove: $Fa \leftrightarrow (\forall x)(x=a \rightarrow Fx)$

*	1	Fa	PREM
**	2	$b=a$	PREM
**	3	Fb	2, 1 LL
*	4	$b=a \rightarrow Fb$	2–3 CP
*	5	$(\forall x)(x=a \rightarrow Fx)$	4 UG (b)
	6	$Fa \rightarrow (\forall x)(x=a \rightarrow Fx)$	1–5 CP
*	7	$(\forall x)(x=a \rightarrow Fx)$	PREM
*	8	$a=a \rightarrow Fa$	7 UI
*	9	$a=a$	BL
*	10	Fa	8, 9 MP
	11	$(\forall x)(x=a \rightarrow Fx) \rightarrow Fa$	7–10 CP
	12	$Fa \leftrightarrow (\forall x)(x=a \rightarrow Fx)$	6,11 ADJ, DEF $\leftrightarrow$
			QED

It is perhaps worth noting here that the deduction from left to right (steps 1 to 5) is quite different from the deduction from right to left (steps 7 to 10). Left-to-right uses Leibniz's law and universal generalization (restricted), with conditional proof; right-to-left uses the law of self-identity (Butler's law), universal instantiation (unrestricted), and modus ponens.

Let us see how an instance of this deduction will read in ordinary language:

*	1	Arthur is forty-two years old	PREM
**	2	Let us assume that Buzz is identical with Arthur – we have given Arthur the nickname "Buzz".	PREM

∗∗	3	By Leibniz's law it follows that Buzz is forty-two years old.	2, 1 LL
∗	4	We conclude (by Conditional Proof) that if Arthur and Buzz "are" identical then Buzz is forty-two years old.	2–3 CP

But 'Buzz' was a nickname picked at random, so we may generalize:

∗	5	Everyone identical with Arthur is forty-two years old.	4 UG (*b*)
	6	If Arthur is forty-two years old then everyone identical with Arthur is forty-two years old.	1–5 CP

Next we establish the converse.

∗	7	Everyone identical with Arthur is forty-two years old.	PREM
∗	8	Accordingly, if Arthur is identical with Arthur then Arthur is forty-two years old.	7 UI
∗	9	But Arthur is identical with Arthur.	BL
∗	10	So Arthur is forty-two years old.	8, 9 MP
	11	We have shown that if everyone identical with Arthur is forty-two years old then Arthur is forty-two years old.	7–10 CP

Putting these two conclusions together, we see that we have established that

	12	Arthur is forty-two years old if and only if everyone identical with Arthur is forty-two years old	6, 11 ADJ, Def ↔ QED

Let us see next what this equivalence shows us about how to formalize "Henry and only Henry is in the room," or "Henry is the one and only person in the room":

$$Rh \land (\forall x)(Rx \to x=h)$$

Since, as we have shown, Rh is equivalent to and so interchangeable with $(\forall x)(x=h \to Rx)$,

$$Rh \land (\forall x)(Rx \to x=h)$$

can be rewritten:

$$(\forall x)(x=h \to Rx) \land (\forall x)(Rx \to x=h)$$

and since $(\forall x)(Fx) \land (\forall x)(Gx)$ is equivalent to $(\forall x)(Fx \land Gx)$ (see the equivalences listed in the Reminders for chapters 4 and 6), the conjunction above may be consolidated:

$$(\forall x)[(x=h \to Rx) \land (Rx \to x=h)]$$

and finally, by the definition of the biconditional,

$$(\forall x)(Rx \leftrightarrow x=h)$$

The statement that Henry is the one and only person in the room can be thought of either as a conjunction – Henry is in the room and nobody else is – or as a universally quantified biconditional, to the effect that, in general, anyone is in the room if and only if he is identical with Henry. This seems conceptually right (the reader may wish to check that observation against his/her own intuitions), and, notationally, it is short and neat. But it does not state explicitly that Henry is in the room, and, in the course of argument, one may wish to recover that bit of information.

We have seen how the biconditional form can be built from the conjunctive form; let us reverse the process and show how to derive from the biconditional formulation the two conjuncts: Henry is in the room, and nobody else is.

The Logic of Identity

*   1 $(\forall x)(Rx \leftrightarrow x=h)$ PREM

              To Prove: Rh

*   2 $Rh \leftrightarrow h=h$ 1 UI
*   3 $h=h$ BL
*   4 $h=h \rightarrow Rh$ 2 DEF $\leftrightarrow$, SIMP
*   5 Rh 4, 3 MP
                                        QED

             To Prove: $\sim(\exists x)(Rx \wedge x \neq h)$

**   6 $(\exists x)(Rx \wedge x \neq h)$ PREM (for RED)
**   7 $Ra \wedge a \neq h$ 6 EI (a)
**   8 $Ra \leftrightarrow a=h$ 1 UI
**   9 $Ra \rightarrow a=h$ 8 DEF $\leftrightarrow$, SIMP
** 10 $\sim(Ra \rightarrow a=h)$ 7 INT (NIF)
** 11 $(Ra \rightarrow a=h) \wedge \sim(Ra \rightarrow a=h)$ 9, 10 ADJ
*   12 $\sim(\exists x)(Rx \wedge x \neq h)$ 6-11 RED
                                        QED

Notice the contrast between step 4 and step 9. In simplifying the biconditionals (steps 2 and 8) we have chosen conditionals going in opposite directions.

Similarly, the statement that there is exactly one person in the room, seen first as the conjunction of two statements:

$$(\exists x)(Rx)$$
There is at least one person in the room.

and

$$(\forall x)(\forall y)[Rx \wedge Ry) \rightarrow x=y]$$
There is at most one person in the room.

can be condensed to a quantified biconditional:

$$(\forall x)(\forall y)(Rx \leftrightarrow x=y)$$
There is exactly one person in the room.

So we have found two ways of formalizing the number 1 – not in the abstract, but in use. We can think of the notion of unity as the conjunction of existence and uniqueness – the mathematician's way – or as the more elegant (but intuitively more elusive) notion of there being something such that having a certain property (here, being in the room) is tantamount to being it. "I have one uncle" means that there is someone such that being my uncle is tantamount to being him. This is false if I have several uncles or if I have none.

These notions can also be used to express the statement that there are two people in the room:

$(\exists x)(\exists y) \{Rx \wedge Ry \wedge x \neq y \wedge (\forall z)[Rz \rightarrow (z=x \vee z=y)]\}$

or

$(\exists x)(\exists y)\{x \neq y \wedge (\forall z)[Rz \leftrightarrow (z=x \vee z=y)]\}$

but the consolidation involved is less interesting in this case.

Truth Trees with Identity

If we are to use truth trees to test the validity of arguments involving identity, we must decide how to introduce the principles of identity into the truth-tree routine. We could systematically introduce instances of BL: $(\forall x)(x=x)$, or $(\forall x)(\forall y)[(x=y \wedge Fx) \rightarrow Fy]$ – LL restated as a theorem rather than as a principle of inference – as extra premises in such arguments, and so use these principles without disturbing the truth-tree routine. But that would be inconvenient and unreliable: it would introduce an element of human ingenuity (remembering that identity theory is relevant to a given argument) into what was intended to be a routine testing procedure. We prefer to build these principles into the truth-tree procedure itself. Accordingly, we shall use them to define two new ways of recognizing inconsistency.

Since everything is identical with itself, it follows that such sentences as '$a \neq a$' and '$b \neq b$' are self-contradictory. Accordingly,

BT: Any statement of the form $x \neq x$ constitutes a contradiction, and so closes any truth-tree branch on which it occurs.

Let us see how this works in practice. To check the validity of $(\forall x)(x=a \rightarrow Fx)$, ∴ Fa, deduced above, we obtain the truth tree:

The Logic of Identity 291

PREM: $(\forall x)(x=a \to Fx)$
~CON: ~Fa

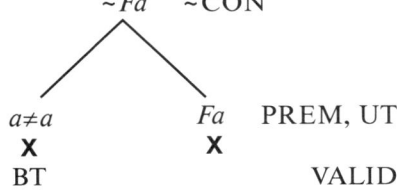

~Fa ~CON

$a \neq a$ Fa PREM, UT
×   ×
BT   VALID

The right-hand branch closes because '~Fa' contradicts 'Fa'; the left-hand branch closes because '$a \neq a$' is self-contradictory, by Butler'r law.

In order similarly to use Leibniz's law (LL, or if you prefer, LL') we turn to its counterclaim. Any instance of

$$Fx \land \sim Fy \land x=y$$

also constitutes a contradiction. We use this observation to test the converse of the argument we have just tested, that is: Fa, $\therefore$ $(\forall x)(x=a \to Fx)$.

PREM: Fa
~CON: ~$(\forall x)(x=a \to Fx)$ $(\exists x)(x=a \land \sim Fx)$ by QN'

        Fa PREM

       $b=a$
       ~Fb ~CON, ET (b)
        ×
       LT   VALID

This pair of truth trees establishes the equivalence:

$$Fa \leftrightarrow (\forall x)(x=a \to Fx)$$

Unfortunately, this initial formulation does not always suffice. In more complicated examples conjunctions of the following form may occur, and these are also (obviously enough) inconsistent by virtue of Leibniz's law:

$$Fx \land \sim Fy \land x=z \land y=z$$

Let us now justify this claim, but postpone discussion of its usefulness. We prove a "stretched" version of Leibniz'z law, the theorem:

$$(Fa \land a=c \land b=c) \to Fb$$

$$* \begin{cases} 1 & Fa \\ 2 & a=c \\ 3 & b=c \end{cases} \quad \text{PREM}$$

* 4 Fc 2, 1 LL
* 5 Fb 3, 4 LL
   6 $(Fa \land a=c \land b=c) \to Fb$ $(1, 2, 3) \to 5$ CP
                                                         QED

Alternatively, we could appeal to the transitivity of '=' ($a=c$ and $c=b$ yield $a=b$) or to the principle (reminiscent of Euclid) that "things identical with the same thing are identical with each other," both of which obviously hold by virtue of the meaning of identity, and which are also easily proved by means of Leibniz's law. One way or another, I trust that the reader will find the following an acceptable addition to the truth-tree rules:

LT: Any conjunction of the form
    $Fx \land \sim Fy \land x=y$
or of the form
    $Fx \land \sim Fy \land x=z \land y=z$
constitutes a contradiction, and so closes any truth-tree branch on which it occurs.

'Fx' and '$\sim Fy$' are of course initially consistent, not contradictory, precisely because 'x' and 'y' may well refer to different things, and so, as we have seen in chapter 5, they do not close a truth-tree branch on which they both appear. But, if they are conjoined either with an identity statement '$x=y$' or with a pair of identity statements that identify both x and y with some "third" entity z, the result is contradictory and so closes the branch in question.

Two points should be noted. First,

$$Fx \land Fy \land x \neq y$$

does *not* constitute a contradiction. This is obvious on the face of it. New York is a city and Paris is a city and New York is not identical with Paris, and there is nothing inconsistent about that. This is worth mentioning only because it is so easy to be careless about *a*s and *b*s and *x*s and *y*s if one forgets about meaning. Denials of identity are *of no use* for Leibniz's law. On a truth-tree branch '$x \neq y$' directly contradicts '$x=y$', and '$x \neq x$' is itself self-contradictory, by Butler's Law, BL; otherwise, the useful items will be affirmations of identity.

Secondly, an identity statement '$a=b$' can be seen as predicating "is identical with a" of b or as predicating "is identical with b" of a. Similarly,

'$a \neq b$' tells us about b that it is other than a and about a that it is other than b.

Accordingly,

$$a=c \land b=c \land a \neq b$$

constitutes a contradiction according to Leibniz's law. One must be on the lookout for contradictions of this kind.

Some Examples

Let us turn to some examples. First, the consequences of the fact that George Eliot is identical with Mary Ann Evans:

(A) P1. Mary Ann Evans was an Englishwoman.
    P2. Mary Ann Evans translated Spinoza's *Ethics* into English.
    P3. George Eliot wrote *Middlemarch*.
    P4. George Eliot was Mary Ann Evans.

From these premises a number of conclusions can be drawn, among them

    C1. George Eliot was an Englishwoman.
    C2. Mary Ann Evans wrote *Middlemarch*.
    C3. An Englishwoman wrote *Middlemarch*.
    C4. Someone was both a translator and a writer.

294   Identity and Description

Let *Ex* = *x* is an Englishwoman
Let *Txy* = *x* translated *y*
Let *Wxy* = *x* wrote *y*
Let *a*   = Mary Ann Evans
Let *g*   = George Eliot
Let *e*   = Spinoza's *Ethics*
Let *m*   = *Middlemarch*

P1. *Ea*   C1. *Eg*
P2. *Tae*   C2. *Wam*
P3. *Wgm*   C3. (∃x)(*Ex* ∧ *Wxm*)
P4. *a=g*   C4. (∃x)[(∃y)(*Txy*) ∧ (∃y)(*Wxy*)]

Let us check out C3 and C4.

~C3: (∀x)(~*Ex* ∨ ~*Wxm*)   by QN, NAND

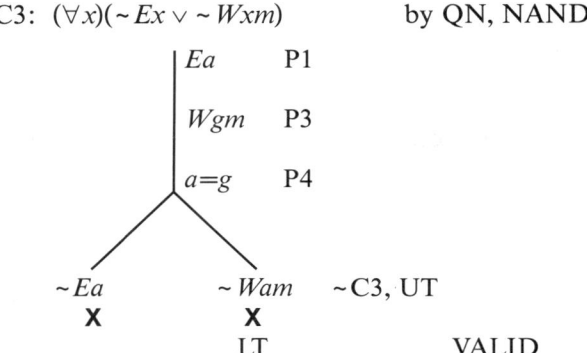

Ea	P1
Wgm	P3
a=g	P4

~*Ea*   ~*Wam*   ~C3, UT
 X   X
            LT   VALID

C4:   ~(∃x)[(∃y)(*Txy*) ∧ (∃z)(*Wxz*)]
       (∀x)[~(∃y)(*Txy*) ∨ ~(∃z)(*Wxz*)]   QN, NAND
       (∀x)[(∀y)(~*Txy*) ∨ (∀x)(~*Wxz*)]   QN
~C4″: (∀y)(~*Tay*) ∨ (∀z)(~*Waz*)   UT
~C4″: (∀y)(~*Tgy*) ∨ (∀z)(~*Wgz*)   UT

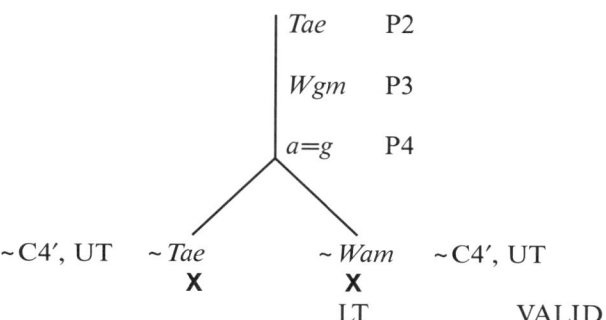

Either ~C4' or ~C4" would suffice to close the tree. Using C4', we used Leibniz's law on the right-hand side; using C4", we would use it on the left.

Next, a small problem in formalization.

(B) If someone telephoned, John did.
   Therefore, if someone telephoned, it was John.

The premise is easy to symbolize. We could write it this way:

$$(\exists x)(Tx) \to Tj$$

or, if we prefer, we could stretch the scope of the quantifier and obtain

$$(\forall x)(Tx \to Tj)$$

Either is correct.

The conclusion presents a familiar problem in cross reference. In "If someone telephoned, it was John" what does the "it" refer to? Surely "it" refers to the "someone" of the antecedent, and surely we will need to use both a wide-scope universal and the concept of identity. Result: $(\forall x)(Tx \to x=j)$.

PREM: $(\exists x)(Tx) \to Tj$
CON: $(\forall x)(Tx \to x=j)$

~CON: $(\exists x)(Tx \land x \neq j)$ by QN'

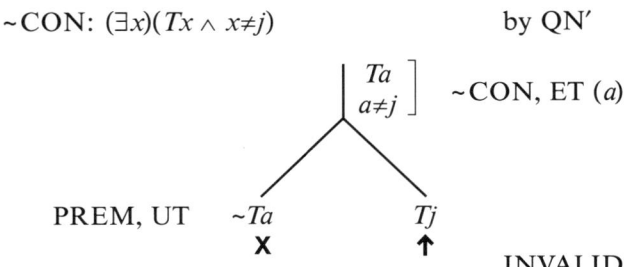

The premise, "If someone telephoned, John did" does not rule out the possibility that someone other than John, say, Algernon, telephoned as well. So this argument turns out to be invalid. A careless reading might have led us to guess that it was not. But the converse is valid.

PREM: $(\forall x)(Tx \to x=j)$
CON: $(\exists x)(Tx) \to Tj$
~CON: $(\exists x)(Tx) \land {\sim}Tj$ by NIF

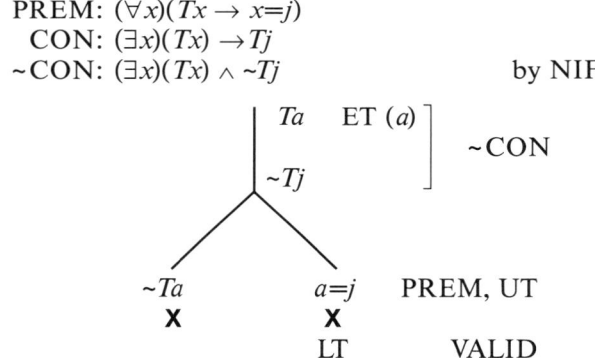

Finally, a slightly more complicated example: two related arguments, one of them obviously invalid and the other valid:

(C) Some people trust themselves but no one else.
    Therefore, there is no one whom everybody trusts.

    There are at least two people who trust themselves but no one else.
    Therefore, there is no one whom everybody trusts.

The first one:

PREM: $(\exists x)[Txx \wedge (\forall y)(Txy \to x=y)]$
$Taa \wedge (\forall y)(Tay \to a=y)$ by ET (a)
CON: $\sim(\exists x)(\forall y)(Tyx)$
~CON: $(\exists x)(\forall y)(Tyx)$ by DN
$(\forall y)(Tyb)$ by ET (b)

```
         | Taa        PREM, ET (a)
         | Tab  ]
         | Tbb  ]    ~CON, ET (b), UT
        / \
     ~Taa   a=a
      X    / \
        ~Tab   a=b  ]  PREM, UT
         X      ↑
                    INVALID
```

It is possible that everyone trusts b, who is identical with a.

The second one:

PREM: $(\exists x)(\exists y)\{Txx \wedge Tyy \wedge x \neq y \wedge (\forall z)[(Txz \to x=z)$
$\wedge (Tyz \to y=z)]\}$
$(\exists y)\{Taa \wedge Tyy \wedge a \neq y \wedge (\forall z)[(Taz \to a=z)$
$\wedge (Tyz \to y=z)]\}$ by ET (a)
$Taa \wedge Tbb \wedge a \neq b \wedge (\forall z)[(Taz \to a=z)$
$\wedge (Tbz \to b=z)]$ by ET(b)
CON: $\sim(\exists x)(\forall y)(Tyz)$
~CON: $(\exists x)(\forall y)(Tyz)$ by DN
$(\forall y)(Tyc)$ by ET(c)

298 Identity and Description

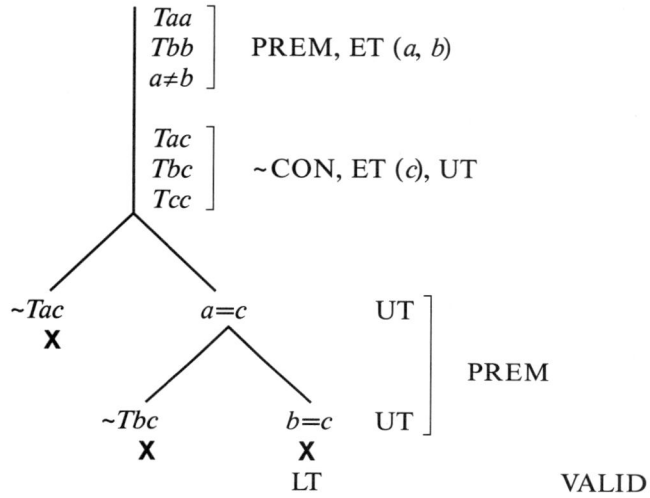

Reminders for Chapter 8

Rules of Inference for Identity

Leibniz's law: Given a statement of identity, its terms are interchangeable within any statement containing either of those terms.

Accordingly,
$$\frac{x=y}{Fx} \quad \text{LL}$$
$$\therefore Fy$$

Leibniz's law': Given a pair of statements, one affirmed, one denied, which are otherwise exactly alike except for the interchange of a pair of terms, it follows that the items denoted by that pair of terms are not identical.

Accordingly,
$$\frac{Fx}{\sim Fy} \quad \text{LL}'$$
$$\therefore x \neq y$$

The Axiom of Self-identity, or Butler's law: Everything is identical with itself.

$$BL: (\forall x)(x=x)$$

This axiom is to be used as if it were a theorem. Accordingly, it may be introduced as a step in proof at any point in any deduction. And any alphabetical variant, say

$$(\forall z)(z=z)$$

or any instance of it, the result of substituting a letter for x, say

$$b=b$$

may likewise be introduced as a step in proof.

Truth-tree Rules for Identity

BT: Any statement of the form

$$x \neq x$$

constitutes a contradiction and, therefore, closes any truth-tree branch on which it occurs.

LT: Any conjunction of the form

$$Fx \land {\sim} Fy \land x=y$$

or of the form

$$Fx \land {\sim} Fy \land x=z \land y=z$$

constitutes a contradiction and, therefore, closes any truth-tree branch on which it occurs.

It is to be noted that statements of *non*identity, or otherness, are useless for Leibniz's law, whether taken as a principle of inference in natural deduction or as a truth-tree rule.

Problems for Chapter 8

I. 1. Check with a pair of truth trees that "Someone trusts some people and distrusts some people" is indeed equivalent to "Someone trusts some people and distrusts others."

2. Do a deduction for:

   At least two people trust themselves but no one else. Therefore, there is no one whom everybody trusts.

3. Show, both by a pair of truth trees and by a pair of deductions, that "Socrates is wise" is equivalent to "Someone who is identical with Socrates is wise."

4. Show, both by a pair of truth trees and by a pair of deductions, that "Someone who is identical with Socrates is wise" is equivalent to "Everyone who is identical with Socrates is wise."

5. Test the following statement by truth tree, and do a deduction if valid:

   If Mary is tolerant of everyone except herself, Mary is tolerant of everyone.

6. Test the following statement by truth tree, and do a deduction if valid:

   If Mary is tolerant of everyone, Mary is tolerant of everyone except herself.

II. Check the following arguments by truth tree and do deductions for the valid ones.

1. Someone who dislikes Mary put poison in her soup.
   Bella has known Mary for many years and likes Mary very much.
   Therefore, whoever put poison in Mary's soup, it was not Bella.

2. Someone who dislikes Mary put poison in her soup.
   Bella has known Mary for years and likes her very much.
   Therefore, someone other than Bella put poison in Mary's soup.

3. Someone who dislikes Mary put poison in her soup.
   (Unlike Bella) Alice and Clara dislike Mary.
   Therefore Alice or Clara must have put poison in Mary's soup.

4. Someone who dislikes Mary put poison in her soup.
   Both Henry and Joseph like Mary very much.
   Therefore, neither Henry nor Joseph put poison in Mary's soup.

5. Someone who dislikes Mary put poison in her soup.
   Everyone likes Mary except Elinor and Dolores.
   Therefore either Elinor or Dolores put poison in Mary's soup.

6. No two (people) like all the same books.
   No one likes a book who hasn't read it.
   Therefore, if there is a book that is read by at least two (people), there is a book that is liked by someone and disliked by someone else.

7. No two (people) like all the same books.
   No one likes a book who hasn't read it.
   Therefore, any book that is read by at least two (people) is liked by someone and disliked by someone else.

8. The one and only (person) whom everybody loves is not the same (person) as the one who loves everybody.
   Therefore, there are unrequited lovers – people who love but are not loved in return.

[Reminder: In dealing with these examples, it is well to consult your intuitions about validity before doing the formal work, and then use the formal work to check the intuitions.]

Chapter 9
Definite Description

The definite article 'the' in English (*der, die, das* in German, *le* or *la* in French, *el* or *la* in Spanish, and so forth) is used for many purposes, but its primary purpose is for what Bertrand Russell called "definite description."[1] In such phrases as "the book you are reading," "the house where I was born," "the proper way to hold a knife for chopping onions," and "the road to Albany," 'the' is roughly synonymous with 'the one and only' and conveys the sense of the number 1. Use of such phrases presupposes both the existence and the uniqueness of the entity described, whereas use of the indefinite article 'a' or 'an' posits only existence. (Interestingly enough, the word for 'a' in the other languages cited is the same word as the word for 'one', although it means only "at least one," not "exactly one.")

Logicians, since the publication of *Principia Mathematica*,[2] have used an inverted Greek iota 'ι' as symbol for "the one and only." I shall try to follow and explain Russell's analysis, but we shall have little need for the iota notation itself. Occasionally the word 'one' in parenthesis will remind the reader that we are dealing with a definite description.

Russell observed that such sentences as "The bus now arriving is overcrowded" (Russell's own favorite example was "The present king of France is bald") make three claims, in this case:

---

[1] Bertrand Russell, "On Denoting," *Mind*, n.s. vol. 14 (1905), pp. 479–93, variously reprinted, notably in *Logic and Knowledge, Essays*, R. C. March, ed. (London: Allen and Unwin, 1956). "Descriptions," chap. 16 of *Introduction to Mathematical Philosophy* (London: Allen and Unwin; New York: Macmillan, 1919), also variously reprinted.

[2] Alfred North Whitehead and Bertrand Russell (Cambridge: University Press, 1910), hereafter, *PM*.

There is a bus now arriving.   (existence)
No other buses are now arriving.   (uniqueness)
This bus is overcrowded.   (whatever)

Accordingly, "The bus now arriving is overcrowded" can be false on any of three grounds: no bus is now arriving; two or more buses are now arriving; (one and only one bus is now arriving, but) it is not overcrowded.

Let us note right away that the sentence "The bus now arriving is not overcrowded" is *not* the negation of the statement we are discussing; for, like that statement, it presupposes the existence and uniqueness of the bus now arriving. The negation of that statement must be expressed in English by some such clumsy locution as "It is not the case that the bus now arriving is overcrowded" -- unless we are in a position to make ourselves clear by just saying "no". This negation consists of the three-pronged disjunction whose members are listed at the end of the last paragraph, i.e.,

No bus is now arriving, or
Two or more buses are now arriving, or
One bus is now arriving, but it is not overcrowded.

"The bus now arriving is not overcrowded" is only one – the third – of those disjuncts. Incidentally, some negative sentences similar in linguistic structure to this one are ambiguous in this respect; this one is not. "The book you need is not in the library," for example, can be taken either way. It can be taken to mean either "There is a specific book that you need, which is not in the library" or (roughly) "None of the books in the library are what you need."

Let us put all this more formally. "The one and only F is G" will be written in the iota notation:

$$G(\iota x)(Fx)$$

which can be read:

G holds of the one and only
entity x such that x is F.

And this in turn will be defined, in the notation of chapter 8:

$$(\exists x)[(\forall y)(Fy \leftrightarrow y=x) \wedge Gx]$$

or

$$(\exists x)[Fx \wedge (\forall y)(Fy \rightarrow y=x) \wedge Gx]$$
There is one and only one F, and it is G.

The first formulation above (Quine's) is more compact and perhaps clearer; for it makes visually obvious the difference in function between 'F' and 'G' in such sentences. After all, "The one and only bus that is now arriving is overcrowded" says something quite different from "The one and only bus that is overcrowded is now arriving," and our notation should remind us of this difference. The second formulation (Russell's), on the other hand, is more convenient for many purposes. The fact that the two formulations are equivalent has already been demonstrated in chapter 8.

Either way, "The F is G" is taken to be an existential claim, written with a wide-scope existential quantifier, whose scope cannot be shrunk. "There is an entity (x) such that it is F and only it is F and it is G." In spite of its complexity, "The F is G" cannot be regarded as a conjunction, ready to be simplified; the conjunction is its open sentence. So simplification is in order only after the entity in question has been "named," that is, after the rule EI has been applied.

Accordingly, when a sentence of the form "The F is G" or "$G(\imath x)(Fx)$" appears as a premise in a deduction, the deduction might begin like this:

```
* 1  (∃x)[(∀y)(Fy ↔ y=x) ∧ Gx]          PREM
* 2  (∀y)(Fy ↔ y=a) ∧ Ga                1 EI (a)
* 3  (∀y)(Fy ↔ y=a)                     2 SIMP
* 4  Ga                                 2 SIMP
```

or like this:

```
* 1  (∃x)[Fx ∧ (∀y)(Fy → y=x) ∧ Gx]     PREM
* 2  Fa ∧ (∀y)(Fy → y=a) ∧ Ga           1 EI (a)
* 3  Fa                                 2 SIMP
* 4  (∀y)(Fy → y=a)                     2 SIMP
* 5  Ga                                 2 SIMP
```

For convenience, we may condense such openings, respectively, in the following way:

$$*\begin{cases} 1 & (\forall y)(Fy \leftrightarrow y=a) \\ 2 & Ga \end{cases} \quad \text{PREM } (a)$$

or

$$*\begin{cases} 1 & Fa \\ 2 & (\forall y)(Fy \rightarrow y=a) \\ 3 & Ga \end{cases} \quad \text{PREM } (a)$$

omitting steps 1 and 2 of the first versions.

Preparation of premises before beginning a deduction is, of course, not new. What is of interest here is the flag, which calls attention to the prior use of the rule of naming. The premise:

$$* \quad 0 \quad G(\mathfrak{i}x)(Fx)$$

which has been defined as an existential, is, so to speak, lurking in the background, and the flag '(a)' reminds us of that fact. It also reminds us that it would be fallacious to generalize from 'a' at some point in the ensuing deduction.

Before going on to deal with some examples, we need to mention one further matter of notation. A recurrent context in which definite descriptions occur is in such sentences as "Joe is the fellow I was telling you about," or "Mary is the president of the class"; that is,

The (one) F is identical with a.

or

$$(\mathfrak{i}x)(Fx)=a$$

Using the definition of '$G(\mathfrak{i}x)(Fx)$' given above, we would obtain

$$(\exists x)[(\forall y)(Fy \leftrightarrow y=x) \land x=a]$$

which is clumsy. But we saw in chapter 8 that "The F is a" can be written

$$(\forall y)(Fy \leftrightarrow y=a)$$

And we also saw in chapter 8 that "Fa" is in general equivalent to "There is something which is F and is identical with a." (The example that we discussed in chapter 8 was the equivalence of

"Socrates is wise" and "There is someone who is wise and is identical with Socrates.")

$$(\exists x)(Fx \wedge x=a) \leftrightarrow Fa$$

An instance of this equivalence – putting $(\forall y)(Fy \leftrightarrow y=x)$ for Fx – is

$$(\exists x)[(\forall y)(Fy \leftrightarrow y=x) \wedge x=a] \leftrightarrow (\forall y)(Fy \leftrightarrow y=a)$$

So we are free to use the shorter formulation of chapter 8, knowing that the two definitions agree. Henceforth we will write

$$(\imath x)(Fx)=a$$
as
$$(\forall x)(Fx \leftrightarrow x=a)$$

Some Deductions and Truth Trees

Here is an example of how a descriptive premise can function in a deduction.

(A) The person I had lunch with yesterday is charming and speaks Swahili.
Therefore, the charming person I had lunch with yesterday speaks Swahili.

$$(\exists x)[(\forall y)(Ly \leftrightarrow y=x) \wedge Cx \wedge Sx]$$
$$\therefore (\exists x)\{(\forall y)[(Cy \wedge Ly) \leftrightarrow y=x] \wedge Sx\}$$

*	1	$(\forall y)(Ly \leftrightarrow y=a)$	
*	2	Ca	PREM (a)
*	3	Sa	
*	4	$La \leftrightarrow a=a$	1 UI
*	5	$a=a$	BL
*	6	$a=a \to La$	4 DEF $\leftrightarrow$, SIMP
*	7	La	6, 5 MP
**	8	$Cb \wedge Lb$	PREM
**	9	Lb	8 SIMP
**	10	$Lb \leftrightarrow b=a$	1 UI
**	11	$Lb \to b=a$	10 DEF $\leftrightarrow$, SIMP
**	12	$b=a$	11, 9 MP

```
 * 13  (Cb ∧ Lb) → b=a                         8–12 CP
** 14  b=a                                     PREM
** 15  Cb                                      14, 2 LL
** 16  Lb                                      14, 7 LL
** 17  Cb ∧ Lb                                 15, 16 ADJ
 * 18  b=a → (Cb ∧ Lb)                         14–17 CP
 * 19  (Cb ∧ Lb) ↔ b=a                         13, 18 ADJ, DEF ↔
 * 20  (∀y)[(Cy ∧ Ly) ↔ y=a]                   19 UG (b)
 * 21  (∀y)[(Cy ∧ Ly) ↔ y=a] ∧ Sa              20, 3 ADJ
 * 22  (∃x){(∀y)[(Cy ∧ Ly) ↔ y=x] ∧ Sx}        21 EG
                                                QED
```

I leave the truth tree to the reader. But let us look at the deduction in some detail. Given the premise, the descriptive phrase 'the person I had lunch with yesterday' succeeds in denoting; that is, the one and only one person I had lunch with yesterday exists [Whitehead and Russell[3] would write this: E! (ɿx)(Lx)]. The major project in deriving the conclusion is to show that the longer descriptive phrase 'the charming person I had lunch with yesterday' likewise succeeds in denoting, that is, that the one and only charming person I had lunch with yesterday "also" exists. (It is obvious that, if such a person exists, he or she speaks Swahili – that part of the deduction is easy; see premise 3.) Premise 1 alone suffices to show that I had lunch with no more than one charming person yesterday (we get as far as step 13 without appeal to any premise except 1), since I had lunch yesterday with no more than one person of any kind. But premise 2 is needed at step 15. In other words, the following argument (omitting premise 2) would be invalid:

The person I had lunch with yesterday speaks Swahili.
Therefore, the charming person I had lunch with yesterday speaks Swahili.

This less informative premise does not warrant the use of the word 'charming' in the conclusion. It is easy to overlook little items like that.

---

[3] *PM*, 1st ed., p. 32; 2d ed., pp. 30, 31.

It is also worth noting that the converse of argument A is invalid:

> The charming person I had lunch with yesterday speaks Swahili. Therefore, the person I had lunch with yesterday is charming and speaks Swahili.

The fact that 'the charming person I had lunch with yesterday' succeeds in denoting does *not* imply that 'the person I had lunch with yesterday' succeeds likewise. Given this premise, I certainly did have lunch with someone yesterday, but I might have had lunch with several people, only one of whom was charming. This we confirm by truth tree:

PREM 1: $(\forall y)[(Cy \wedge Ly) \leftrightarrow y=a]$
PREM 2: Sa
~CON: ~$(\exists x)[(\forall y)(Ly \leftrightarrow y=x) \wedge Cx \wedge Sx]$
  $(\forall x)$~$[(\forall y)(Ly \leftrightarrow y=x) \wedge Cx \wedge Sx]$ by QN
  $(\forall x)[$~$(\forall y)(Ly \leftrightarrow y=x) \vee $~$Cx \vee $~$Sx]$ by NAND
  $(\forall x)[(\exists y)$~$(Ly \leftrightarrow y=x) \vee $~$Cx \vee $~$Sx]$ by QN
  $(\forall x)[(\exists y)(Ly \leftrightarrow y \neq x) \vee $~$Cx \vee $~$Sx]$ by NIFF
  $(\exists y)(Ly \leftrightarrow y \neq a) \vee $~$Ca \vee $~$Sa$ by UT

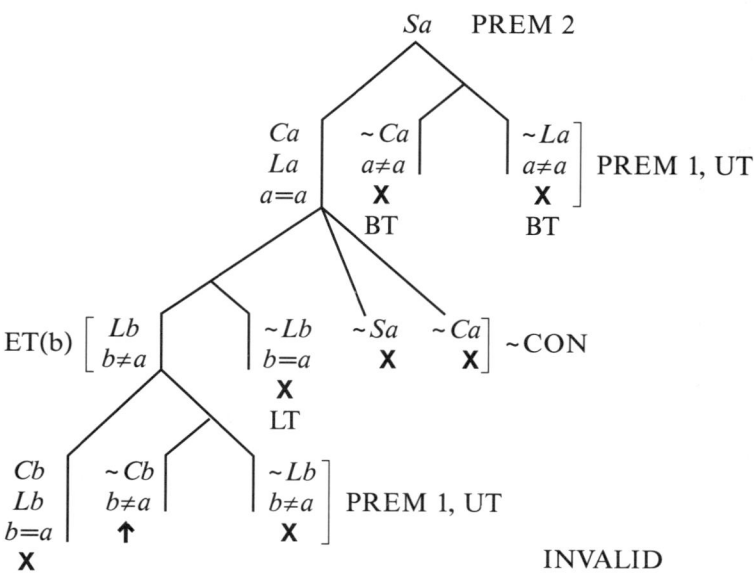

INVALID

The open path represents a situation in which I had lunch with two people: Arnold (*a*) and Bella (*b*); Arnold was charming and spoke Swahili; Bella was not charming.

Next, a slightly more complicated example (intended to be instructive in more than one way, that is, to teach spelling as well as logic).

> (B) The correct spelling of "argument" is with only one 'e'.
> 'Arguement' is a way of spelling "argument" with more than one 'e'.
> Therefore, 'arguement' is incorrect.

As we have seen in chapter 3, the propositional connective 'only if' is unambiguous in English; it does not mean "if and only if." One can make mistakes, of course, but both phrases are nonetheless quite clear. When we come to the "only" of identity theory, however, the situation is more complex. "Only one" may very well mean "precisely one" rather than just "no more than one"; and we ordinarily rely upon context to discover which meaning was intended. This argument is a case in point.

Surely, the first premise tells us both that there is one *and* only one correct spelling of "argument," and that that spelling contains one *and* only one 'e'. It would be misleading to formulate it otherwise.

All the same, the sense of the argument depends upon uniqueness, not upon existence. The "core" argument, which we sketch in our heads when assessing the validity of argument B, before we do the formal work, sets aside the "one and" in both cases and focuses on the "only." The following core argument is intuitively valid:

> B': There is no more than one correct spelling of "argument" (call it *c*); *c* contains no more than one 'e'.
> 'Arguement' is a way of spelling "argument"; 'arguement' contains more than one 'e'.
> Therefore, 'arguement' is incorrect.

A sketch of a deduction for B' would involve only four steps. From the second clauses of both premises, by Leibniz's law', LL', it follows that 'arguement' is not identical with *c*. Instantiating the first clause of the first premise, taken as a universal, we find that if 'arguement' is a correct spelling of "argument" it is identical with c. But it is not. So, by modus tollens, 'arguement' is not a correct spelling of "argument." The first clause of the second premise reminds us that 'arguement' is *a* spelling of "argument"; so, by nonconjunction, it follows that 'arguement' is not correct. QED.

One further comment about usage. As we have seen, "only if" as a sentential connective:

I will go only if I am invited.

and, accordingly, "only" in the plural:

Only fools drive when they have been drinking.

are unambiguously one-directional. But "only" in the singular is ambiguous, sometimes intended to have existential import and sometimes not. In contrast to both of these, "only" in the *negative* singular is (ordinarily at least) unambiguous and compound.

Mary is not the only talented student in the class.

tells us both that Mary is a talented student in the class and that there are others.

Let us return to argument B, as it stands.

Let Sxy = x is a way of spelling y.
Let Exy = x is a letter 'e' in y.
Let Cx = x is correct.
Let a = "argument" (the English word).
Let b = 'arguement' (the spelling).

Accordingly,

x is the correct spelling of "argument"

will be written

$$(\forall y)[(Sya \wedge Cy) \leftrightarrow y=x]$$

and

x has only one 'e'

will be written

$$(\forall z)(\forall w)[(Ezx \wedge Ewx) \leftrightarrow z=w]$$

Definite Description

PREM 1: $(\exists x)\{(\forall y)[(Sya \wedge Cy) \leftrightarrow x=y]$
$(\forall z)(\forall w)[(Ezx \wedge Ewx) \leftrightarrow z=w]\}$
PREM 2: $Sba \wedge (\exists z)(\exists w)(Ezb \wedge Ewb \wedge z \neq w)$

To Prove: $\sim Cb$

The first premise (naming c beforehand) yields:

PREM 1A: $(\forall y)[(Sya \wedge Cy) \leftrightarrow c=y]$
PREM 1B: $(\forall z)(\forall w)[(Ezc \wedge Ewc) \leftrightarrow z=w]$

and the second premise yields:

PREM 2A: Sba
PREM 2B: $(\exists z)(\exists w)(Ezb \wedge Ewb \wedge z \neq w)$
$\sim$CON: Cb by DN

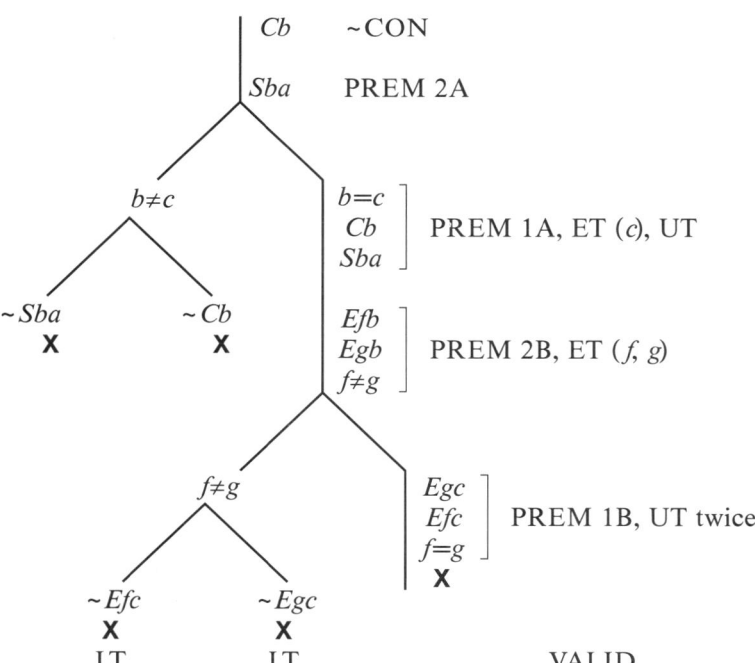

The tree closes, and the argument is valid.

We follow through with a deduction.

$$
*\begin{cases} 1 & (\forall y)[(Cy \land Sya) \leftrightarrow y=c] \\ 2 & (\forall z)(\forall w)[(Ezc \land Ewc) \leftrightarrow z=w)] \\ 3 & Sba \\ 4 & (\exists z)(\exists w)(Ezb \land Ewb \land z \neq w) \end{cases} \quad \text{PREM } (c)
$$

* 5	$(\exists w)(Edb \land Ewb \land d \neq w)$	4 EI (d)
* 6	$Edb \land Egb \land d \neq g$	5 EI (g)
* 7	$(\forall w)[(Edc \land Ewc) \leftrightarrow d=w]$	2 UI
* 8	$(Edc \land Egc) \leftrightarrow d=g$	7 UI
* 9	$(Edc \land Egc) \to d=g$	8 DEF $\leftrightarrow$, SIMP
* 10	$\sim(Edc \land Egc \land d \neq g)$	9 INT, NIF
* 11	$b \neq c$	6, 10 LL'
* 12	$(Cb \land Sba) \leftrightarrow b=c$	1 UI
* 13	$(Cb \land Sba) \to b=c$	12 DEF$\leftrightarrow$, SIMP
* 14	$\sim(Cb \land Sba)$	13, 11 MT
* 15	$\sim Cb$	14, 3 NCONJ
		QED

As we have seen, the deduction for the "core" argument is a good deal simpler:

$$
*\begin{cases} 1 & (\forall y)[(Cy \land Sya) \to y=c] \\ 2 & (\forall z)(\forall w)[(Ezc \land Ewc) \to z=w)] \\ 3 & Sba \\ 4 & (\exists z)(\exists w)(Ezb \land Ewb \land z \neq w) \end{cases} \quad \text{PREM } (c)
$$

* 5	$\sim(\exists z)\sim(\forall w)[(Ezc \land Ewc) \to z=w]$	2 QN
* 6	$\sim(\exists z)(\exists w)(Ezc \land Ewc \land z \neq w)$	5 QN'
* 7	$b \neq c$	4, 6 LL'
* 8	$(Cb \land Sba) \to b=c$	1 UI
* 9	$\sim(Cb \land Sba)$	8, 7 MT
* 10	$\sim Cb$	9, 3 NCONJ
		QED

We may be reminded here of the strategy, suggested in chapter 3 (see page 125), for the preliminary assessment of an argument: strip away inessentials to get at a "core argument," and assess that. This strategy is, I think, helpful in this case.

(C) The bus that is now arriving is late.
   The bus that is now arriving is overcrowded.
   Therefore, the bus that is now arriving is late and overcrowded.

$*\begin{cases} 1 & (\exists x)[(\forall y)(By \leftrightarrow y=x) \wedge Lx] \\ 2 & (\exists x)[(\forall y)(By \leftrightarrow y=x) \wedge Cx] \end{cases}$ PREM

To Prove: $(\exists x)[(\forall y)(By \leftrightarrow y=x) \wedge Lx \wedge Cx]$

*	3 $(\forall y)(By \leftrightarrow y=a) \wedge La$	1 EI (a)
*	4 $(\forall y)(By \leftrightarrow y=b) \wedge Cb$	2 EI (b)
*	5 $(\forall y)(By \leftrightarrow y=a)$	3 SIMP
*	6 $Ba \leftrightarrow a=a$	5 UI
*	7 $a=a$	BL
*	8 $a=a \rightarrow Ba$	6 DEF $\leftrightarrow$, SIMP
*	9 Ba	8, 7 MP
*	10 $(\forall y)(By \leftrightarrow y=b)$	4 SIMP
*	11 $Ba \leftrightarrow a=b$	10 UI
*	12 $Ba \rightarrow a=b$	11 DEF $\leftrightarrow$, SIMP
*	13 $a=b$	12, 9 MP
*	14 Cb	4 SIMP
*	15 Ca	14, 13 LL
*	16 $(\forall y)(By \leftrightarrow y=a) \wedge La \wedge Ca$	3, 15 ADJ
*	17 $(\exists x)[(\forall y)(By \leftrightarrow y=x) \wedge Lx \wedge Cx]$	16 EG
		QED

We turn now to the truth tree.

PREM 1: $(\forall y)(By \leftrightarrow y=a)$ by ET (a)
          La and SIMP
PREM 2: $(\forall y)(By \leftrightarrow y=b)$ by ET (b)
          Cb and SIMP
~CON: $(\forall x)[(\exists y)(By \leftrightarrow y \neq x) \vee \sim Lx \vee \sim Cx]$
                                                  by QN

The diagram is clumsy, but it is manageable.

314   Identity and Description

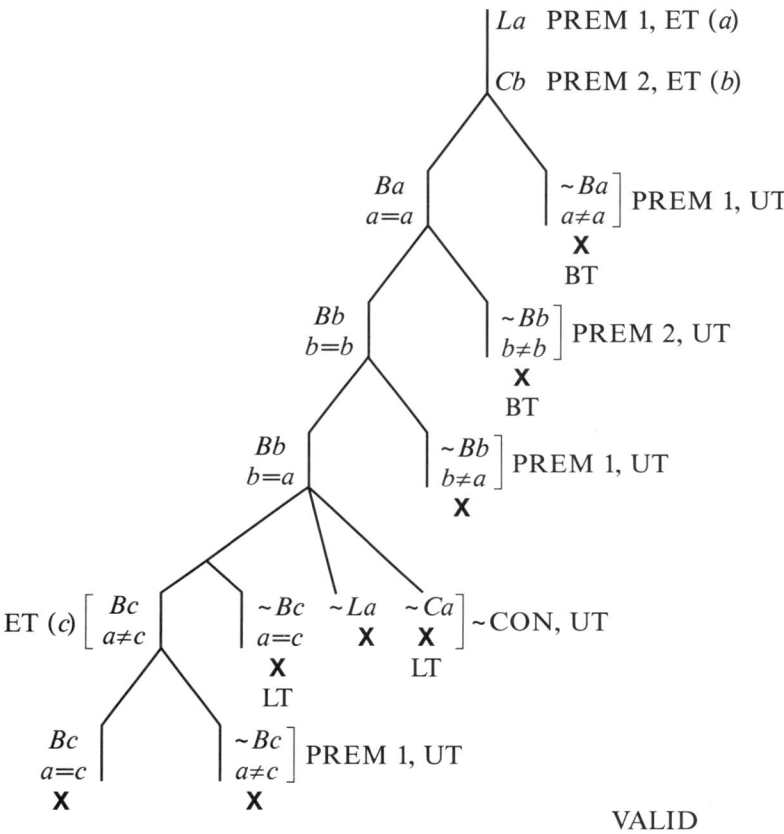

This seems a lot of trouble to go through, to prove something so obviously valid. But it illustrates the pervasiveness of reasoning by principles of definite description in our ordinary thinking, and also the power of the concept of identity. After all, the simpler argument of which this is an instance:

$$(\exists x)(Fx \wedge Gx)$$
$$\underline{(\exists x)(Fx \wedge Hx)}$$
$$\therefore (\exists x)(Fx \wedge Gx \wedge Hx)$$

where the 'F' is not a definite description, is notoriously invalid (see chapters 4 and 5).

The argument given above establishes the difficult half (from left to right) of a useful equivalence, which can be abbreviated (in "bastard" notation):

$$[G(\imath x)(Fx) \wedge H(\imath x)(Fx)] \leftrightarrow (G \wedge H)(\imath x)(Fx)$$

The implication from right to left is easily established by the principles of chapter 4, since the following one-way implication is valid:

$$(\exists x)(Fx \wedge Gx \wedge Hx) \rightarrow [(\exists x)(Fx \wedge Gx) \wedge (\exists x)(Fx \wedge Hx)]$$

The equivalence (abbreviated above):

$$\{(\exists x)[(\forall y)(Fy \leftrightarrow y=x) \wedge Gx] \wedge (\exists x)[(\forall y)(Fy \leftrightarrow y=x) \wedge Hx]\}$$
$$\leftrightarrow (\exists x)[(\forall y)(Fy \leftrightarrow y=x) \wedge Gx \wedge Hx]$$

tells us that our practice of interchanging "The F is G and the F is H" (a conjunction) with "The F is G and H" (an existential) is logically justified.

On Existential Import

The reader will have realized by now that estimating the validity of arguments that involve definite descriptions depends, first, on recognizing a definite description when one sees one and, second, on careful analysis of what each such description amounts to; in this project formulation in symbols is often helpful, even without further formal work.

The following argument, for example, may at first seem valid:

> All lawyers go to law school.
> Therefore, the lawyer who freed Sacco and Vanzetti went to law school.

But it is patently invalid, since its premise is true, but its conclusion is false: Sacco and Vanzetti were executed in August of 1927; nobody freed them. The careless reading that suggested validity at first can be avoided if we formalize:

$$\frac{(\forall x)(Lx \to Sx)}{\therefore \ (\exists x)\{(\forall y)[(Ly \wedge Fy) \leftrightarrow y=x] \wedge Sx\}}$$

Now even a superficial look at this formal restatement should indicate that the argument is invalid. For there is nothing in the premise to suggest what is required for the first clause of the open sentence of the conclusion, namely that one and only one lawyer freed Sacco and Vanzetti. The premise does not guarantee the existence of even one lawyer, let alone that of "the" lawyer who freed Sacco and Vanzetti.

The point being made is that descriptional sentences, sentences of the form:

$$(\exists x)[(\forall y)[(Fy \leftrightarrow y=x) \wedge Gx]$$

make strong existential claims – have strong existential import – whereas denials of such claims and other universal statements do not. Close attention to the existential import of their premises and conclusions is helpful in assessing the validity or invalidity of arguments involving descriptions.

It may be well to recall the argument about the Spartans which we discussed at the end of chapter 5 (argument N). In that case an illusion of validity was generated by an existential assumption: there must be some Spartans, or we would not be talking about them. Here an illusion of validity is created by an even stronger existential assumption: that the (one) lawyer who freed Sacco and Vanzetti existed. And the unreasonableness of making that assumption is obvious the moment that its use is brought to our attention.

What the premise of this argument does imply is the conditional:

> If there was one and only one lawyer who freed Sacco and Vanzetti, then the lawyer who freed Sacco and Vanzetti went to law school.

$$(\exists x)(\forall y)[(Ly \wedge Fy) \leftrightarrow y=x] \to (\exists x)\{(\forall y)[(Ly \wedge Fy) \leftrightarrow y=x] \wedge Sx\}$$

In the iota notation:

$$E! \ (\iota x)(Lx \wedge Fx) \to S(\iota x)(Lx \wedge Fx)$$

Or, more succinctly,

Definite Description 317

If there was one and only one lawyer who freed Sacco and Vanzetti, that lawyer went to a law school.

$$(\forall x)\{(\forall y)[(Ly \wedge Fy) \leftrightarrow y=x] \to Sx\}$$

The reader may wish to establish for him/herself the equivalence of those two formulations. I will discuss instead a similar but simpler case.

At the end of chapter 8 (example B) we formalized

If someone telephoned it was John.

as a universally quantified conditional, in prenex form:

$$(\forall x)(Tx \to x=j)$$

interpreting the 'it' as a pronoun referring back to the "someone" who telephoned. We could, however, read that sentence as saying:

If someone telephoned, the one who telephoned was John.

$$(\exists x)(Tx) \to [(\imath x)(Tx)=j]$$
$$(\exists x)(Tx) \to (\forall y)(Ty \leftrightarrow y=j)$$

Is this reading equivalent to '$(\forall x)(Tx \to x=j)$'?
We use two truth trees to check.

PREM: $(\forall x)(Tx \to x=j)$
~CON: $\sim[(\exists x)(Tx) \to (\forall y)(Ty \leftrightarrow y=j)]$
       $(\exists x)(Tx) \wedge \sim(\forall y)(Ty \leftrightarrow y=j)$ by NIF
       $(\exists x)(Tx) \wedge (\exists y)\sim(Ty \leftrightarrow y=j)$ by QN
       $(\exists x)(Tx) \wedge (\exists y)(Ty \leftrightarrow y \neq j)$ by NIFF

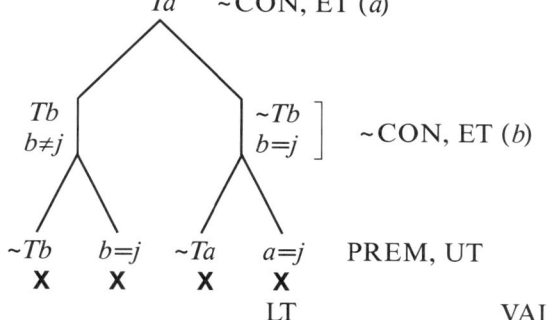

VALID

Note that the right-hand branch uses the "stretched" version of LT.

PREM: $(\exists x)(Tx) \rightarrow (\forall y)(Ty \leftrightarrow y=j)$
$\qquad (\forall x)(\sim Tx) \vee (\forall y)(Ty \leftrightarrow y=j)$ by IF, QN
~CON: $\sim(\forall x)(Tx \rightarrow x=j)$
$\qquad (\exists x)(Tx \wedge x \neq j)$ by QN

$$\begin{array}{c} Ta \\ a \neq j \end{array} \quad \sim\text{CON, ET }(a)$$

PREM, UT $\sim Ta$ $\begin{array}{c} Ta \\ a=j \end{array}$ $\begin{array}{c} \sim Ta \\ a \neq j \end{array}$ PREM, UT
               X   X   X

                                                                        VALID

Like argument B of chapter 8, the following argument would, accordingly, be fallacious:

> If someone telephoned, John did.
> Therefore, if someone telephoned, the one who telephoned was John.

Someone who used such as argument or judged that it was valid would in effect be using the (unjustified) assumption that no more than one person telephoned.

This is an easy mistake to make. An even easier one is to assume that the following is a theorem:

$$(\imath x)(Fx) = (\imath x)(Fx)$$

on the grounds that it is an instance of the axiom of identity, BL, or $(\forall x)(x=x)$. But it is not. What it means is either

> There is one and only one F, and it is identical with the one and only F.

$$(\exists x)[(\forall y)(Fy \leftrightarrow y=x) \wedge x=(\imath y)(Fy)]$$
$$(\exists x)[(\forall y)(Fy \leftrightarrow y=x) \wedge (\forall y)(Fy \leftrightarrow y=x)]$$

or else

The one and only F is self-identical.
$$(\exists x)[(\forall y)(Fy \leftrightarrow y=x) \wedge x=x]$$
both of which are obviously and demonstrably equivalent to
$$(\exists x)(\forall y)(Fy \leftrightarrow y=x)$$
There is one and only one F.

or "E! $(\imath x)(Fx)$," which is decidedly not a theorem. If, indeed, there is one and only one entity that is F, then that entity is self-identical and "$(\imath x)(Fx) = (\imath x)(Fx)$" is true. But that identity statement, which may have appeared to be valid, is *not* independent of the statement of existence; it is equivalent to it; so it is not valid.

As Whitehead and Russell put it, "If $(\imath x)(\phi x)$ has any property whatever, it must exist," and "when $(\imath x)(\phi x)$ exists, it has any property which belongs to everything," including the property of being identical with itself.[4]

On Negation

Finally, I want to discuss briefly the problem of formalizing negative statements that involve definite descriptions. Many such statements are hopelessly ambiguous, as between two forms: on the one hand,

$$\sim G(\imath x)(Fx)$$
$$\sim(\exists x)[(\forall y)(Fy \leftrightarrow y=x) \wedge Gx]$$
It is not the case that the (one) F is G.

which is the negation of

$$G(\imath x)(Fx)$$
$$(\exists x)[(\forall y)(Fy \leftrightarrow y=x) \wedge Gx]$$
The F is G.

and, on the other hand,

---

[4] *PM*, 2d. ed., pp. 174, 175; propositions *14.21, *14.18, *14.28.

$$\bar{G}(\imath x)(Fx)$$
$$(\exists x)[(\forall y)(Fy \leftrightarrow y=x) \wedge \sim Gx]$$
The (one) F is not G.

In the first case the whole descriptive statement is negated; in the second case, only the predicate (that is the force of the overbar, which is here introduced, briefly, for convenience, but which is common in the literature).

The following sentences seem to me to be unambiguous sentences of the second kind:

> The bus that is now arriving is not overcrowded.
>
> The man in the corner is not having a good time.
>
> I do not know where I put my house key.

Each of them conveys the claim that its descriptive phrase succeeds in denoting, that is, respectively, that one and only one bus is now arriving, that there is one and only one man in the corner, and that I left my housekey in one and only one place. So they could be rewritten:

$$(\exists x)\{(\forall y)[(By \wedge Ay) \leftrightarrow y=x] \wedge \sim Ox\}$$
$$(\exists x)\{(\forall y)[(My \wedge Cy) \leftrightarrow y=x] \wedge \sim Gx\}$$

and, letting $Lx = x$ is a location, $Pxy =$ I put x at y, $Kx =$ I know x, and $h =$ my house key,

$$(\exists x)\{(\forall y)[(Ly \wedge Phy) \leftrightarrow y=x] \wedge \sim Kx\}$$

The following, in contrast, seem to me to be unambiguous sentences of the first kind:

> The man who can solve that problem is not stupid.
>
> The house that will please you is not available at your price.
>
> I do not know the man who can beat my brother.
>
> I do not know the answer to your question.

These sentences, unlike the other three, do not have comparable existential import. The first of them seems to mean little more than "No one who can solve that problem is stupid," or

$\sim(\exists x)(Sxp \wedge Tx)$. Similarly for the third one, which probably is intended to convey the boast that no one can beat my brother, or $\sim(\exists x)(Mx \wedge Lxb)$. The second sentence and the last are quite noncommittal. Maybe there are many houses that will please you; maybe there are none. Maybe there are many answers to your question, maybe there are none. Accordingly, all four are best symbolized as denials of existence, say:

$$\sim(\exists x)[(\forall y)[(Syp \leftrightarrow y=x) \wedge Tx]$$
$$\sim(\exists x)\{(\forall y)[(Hy \wedge Py) \leftrightarrow y=x] \wedge Ax\}$$
$$\sim(\exists x)\{(\forall y)[(My \wedge Lyb) \leftrightarrow y=x] \wedge Kx\}$$
$$\sim(\exists x)[(\forall y)[(Ayq \leftrightarrow y=x) \wedge Kx]$$

To take another example – a familiar one –, early in November, 1963, the sentence

The (current) president of the United States likes modern art.
$$L\,(\imath x)(Pxu)m$$
$$(\exists x)[(\forall y)(Pyu \leftrightarrow y=x) \wedge Lxm]$$

was true. John Kennedy was president of the United States, and John Kennedy liked modern art.

On November 22, 1963, John Kennedy was shot, and, for a few hours, between Kennedy's death and the swearing in of Lyndon Johnson, on an airplane in Dallas, there was no president of the United States. Accordingly, the sentence above was false, and so also was:

The president of the United States does not like modern art.

interpreted, as most of us would interpret it, with existential import:

$$(\exists x)[(\forall y)(Pyu \leftrightarrow x=y) \wedge \sim Lxm]$$

even though

$$\sim(\exists x)[(\forall y)(Pyu \leftrightarrow x=y) \wedge Lxm]$$

was true.

On November 23 and thereafter for some time, while Lyndon Johnson was president, both of these last two sentences were true. Lyndon Johnson did not like modern art.

Many negative descriptional sentences are ambiguous; one cannot decide how best to interpret them on the basis of linguistic clues alone. Nonetheless, within any context of argument, it is important to be sensitive to these problems of interpretation and to hold consistently to the interpretive decisions that one makes.

It is useful to remember that the following argument is valid – and that its converse is not:

$$(\exists x)[(\forall y)(Fy \leftrightarrow y=x) \wedge \sim Gx]$$
The one and only F is not G.
$$\therefore \sim (\exists x)[(\forall y)(Fy \leftrightarrow y=x) \wedge Gx]$$
It is not the case that the (one) F is G.

[The first of these statements does imply the second, even though the following two statements, of which the above, respectively, are instances:

$$(\exists x)(Fx \wedge \sim Gx)$$
$$\sim (\exists x)(Fx \wedge Gx)$$

are quite independent.]

We show this validity by deduction:

*	1	$(\exists x)[(\forall y)(Fy \leftrightarrow y=x) \wedge \sim Gx]$	PREM
**	2	$(\exists x)[(\forall y)(Fy \leftrightarrow y=x) \wedge Gx]$	PREM

Using the theorem on conjunction proved above, we consolidate the conjunction of these two premises:

**	3	$(\exists x)[(\forall y)(Fy \leftrightarrow y=x) \wedge Gx \wedge \sim Gx]$	1, 2 ADJ and Theorem
**	4	$(\forall y)(Fy \leftrightarrow y=a) \wedge Ga \wedge \sim Ga$	3 EI (a)
**	5	$Ga \wedge \sim Ga$	4 SIMP
*	6	$\sim (\exists x)[(\forall y)(Fy \leftrightarrow y=x) \wedge Gx]$	2–5 RED QED

We confirm this result by truth tree.

Definite Description

PREM: $(\exists x)[(\forall y)(Fy \leftrightarrow y=x) \wedge \sim Gx]$
  $(\forall y)(Fy \leftrightarrow y=a) \wedge \sim Ga$ by ET (a)
$\sim$CON: $(\exists x)[(\forall y)(Fy \leftrightarrow y=x) \wedge Gx]$ by DN
  $(\forall y)(Fy \leftrightarrow y=b) \wedge Gb$ by ET (b)

The truth tree will diagram:
PREM A: $(\forall y)(Fy \leftrightarrow y=a)$
PREM B: $\sim Ga$
$\sim$CON A: $(\forall y)(Fy \leftrightarrow y=b)$
$\sim$CON B: Gb

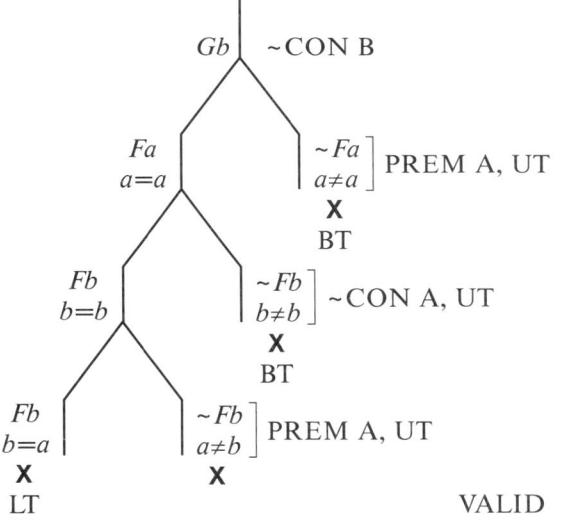

We finish by remarking on the invalidity of the converse:

$$\frac{\sim(\exists x)[(\forall y)(Fy \leftrightarrow y=x) \wedge Gx]}{\therefore (\exists x)[(\forall y)(Fy \leftrightarrow y=x) \wedge \sim Gx]}$$

Its truth tree would be endless, since both the premise and the denial of the conclusion are $\forall\exists$. But it is obvious that the existential claim made in the conclusion: that there is one and only one F, can in no way be warranted by the (universal) premise.

Reminders for Chapter 9

Two definitions:

(1) The F is G.
The one and only F is G.
$G(\iota x)(Fx)$
$(\exists x)[(\forall y)(Fy \leftrightarrow y=x) \wedge Gx]$ (Quine's way)
or
$(\exists x)[Fx \wedge (\forall y)(Fy \rightarrow y=x) \wedge Gx]$ (Russell's way)

(2) x is the F
x is the one and only F
$x = (\iota y)(Fy)$
$(\forall y)(Fy \leftrightarrow y=x)$
or
$Fx \wedge (\forall y)(Fy \rightarrow y=x)$

Theorems with Identity

First, some equivalences established in chapter 8:

$$Fx \leftrightarrow (\exists y)(Fy \wedge y=x)$$

$$Fx \leftrightarrow (\forall y)(y=x \rightarrow Fy)$$

$$(\exists y)[(Fy \wedge y=x) \leftrightarrow (\forall y)(y=x \rightarrow Fy)$$

and, accordingly, the equivalence of the definitions cited above:

$$(\forall y)(Fy \leftrightarrow y=x) \leftrightarrow [Fx \wedge (\forall y)(Fy \rightarrow y=x)]$$

$(\exists x)[(\forall y)(Fy \leftrightarrow y=x) \wedge Gx] \leftrightarrow (\exists x)[Fx \wedge (\forall y)(Fy \rightarrow y=x) \wedge Gx]$

Next, on definite description:

1. If x is the F then x is F.
 $(\forall y)[(Fy \leftrightarrow y=x) \rightarrow Fx$

2. If x is the F then there is no other F.
 $(\forall y)[(Fy \leftrightarrow y=x) \rightarrow \sim(\exists y)(Fy \wedge y \neq x)$

3. If the F is G then there is at least one F.
 $(\exists x)[(\forall y)(Fy \wedge y=x) \wedge Gx] \rightarrow (\exists x)(Fx)$

4. If the F is G then there is no more than one F.
$(\exists x)[(\forall y)(Fy \leftrightarrow y=x) \wedge Gx] \to \sim(\exists x)(\exists y)(Fx \wedge Fy \wedge x \neq y)$

5. If the F is G then some Gs are F.
$(\exists x)[(\forall y)(Fy \leftrightarrow y=x) \wedge Gx] \to (\exists x)(Gx \wedge Fx)$

6. If the F is G and the G is F then "they" are identical.
$(\forall x)(\forall y)\{[(\forall z)(Fz \leftrightarrow z=x) \wedge Gx \wedge (\forall z)(Gz \leftrightarrow z=y) \wedge Fy]$
$\to x=y\}$

7. The F is G and the F is H if and only if the F is G and H.
$\{(\exists x)[(\forall y)(Fy \leftrightarrow y=x) \wedge Gx] \wedge (\exists x)[(\forall y)(Fy \wedge y=x) \wedge Hx]\}$
$\leftrightarrow (\exists x)[(\forall y)(Fy \wedge y=x) \wedge Gx \wedge Hx]$

8. If the F is not-G, then it is not the case that the F is G.
$(\exists x)[(\forall y)(Fy \leftrightarrow y=x) \wedge \sim Gx]$
$\to \sim(\exists x)[(\forall y)(Fy \leftrightarrow y=x) \wedge Gx]$

or, in the iota notation,

$\bar{G}(\imath x) \, Fx \to \sim G(\imath x) \, Fx$

(but *not* vice versa).
See also Reminders for Chapter 8.

Problems for Chapter 9

I. Establish theorems 1 to 6 from the list of theorems above, both by truth tree and by deduction. In the course of your work, use Russell's definition of "the one and only F" at least once and Quine's definition at least once. It will be appropriate throughout to use '*a*'s and '*b*'s (pseudonames rather than free variables) in place of the '*x*'s and '*y*'s used in stating the theorems.

Theorem 7 is established in the text and used in establishing theorem 8. For theorem 8 do a deduction only, *without* appeal to theorem 7.

II. 1. Check the validity of argument (A) in the text, by truth tree:

The person I had lunch with yesterday is charming and speaks Swahili.
Therefore, the charming person I had lunch with yesterday speaks Swahili.

2. Show by deduction that "If someone telephoned it was John" is equivalent to "If someone telephoned, the person who telephoned was John."

3. Explain informally why "The author of 'On Denoting' wrote *Principia Mathematica*" is ordinarily taken to be true, whereas "The author of *Principia Mathematica* wrote 'On Denoting'" is false. (The facts of the case can be found in the footnotes of this chapter if you do not remember them.)

III. Consider the following arguments. In each case, state the argument in symbolic form and decide whether or not it is valid. Then support your judgment, if possible by deduction or by truth tree; if not, by a clear and careful explanation.

1. The man at the window is talking to Eleanor. Therefore, the man who is talking to Eleanor is at the window.

2. The pianist who accompanies Stephen accompanies only talented musicians. Therefore, Stephen is a talented musician.

3. The daring young man on the flying trapeze is a false-hearted lover. Therefore, some lovers are false-hearted.

4. The only student in the class who flunked the swimming test is afraid of the water. Some of the students in the class are not afraid of the water. Therefore, some one in the class did not flunk the swimming test.

5. The man standing at the window is the doctor who operated on Alice. Therefore the doctor who operated on Alice is standing at the window.

6. There is no one standing at the window. Therefore, the doctor who operated on Alice is not standing at the window.

7. The pianist who accompanies Stephen also accompanies Teresa. Therefore, the pianist who accompanies Teresa accompanies Stephen.

8. There is no answer to that question. Therefore, I do not know the answer to that question.

9. Mary is standing at the window with a glass of wine in her hand. Alice is standing at the window with a sandwich in her hand. Therefore, more than one person is standing at the window.

10. George is sitting at the corner table drinking a martini. Henry is not drinking, but he is also sitting at the corner table. Therefore, there is more than one person sitting at the corner table.

11. The doctor who operated on Alice is very disagreeable. Therefore, all the doctors who operated on Alice are very disagreeable.

12. All doctors are disagreeable.
Therefore, the doctor who operated on Alice is disagreeable.

13. The student in the class who flunked the swimming test is afraid of the water.
Some students in the class are not afraid of the water. Therefore, there are at least two students in the class.

14. The woman who stole the violin does not like music. Therefore, some women do not like music.

IV. "George is the most interesting man I know" has been interpreted to mean":

(A) I know George, and he is more interesting than any other man I know.

(B) I know George, and no man I know is more interesting than George is.

(C) George is the one and only man I know who is more interesting than any other man I know.

(D) George is the one and only man I know such that no man I know is more interesting than he is.

Also, by way of background information, we ordinarily assume:

(E) No one is more interesting than anyone who is more interesting than he is.

and, accordingly,

(F) No one is more interesting than himself.

Also,

(G) Anyone who is more interesting than someone who in turn is more interesting than somebody is likewise more interesting than that somebody.

(H) Anyone who is not more interesting than someone who in turn is not more interesting than somebody is not more interesting than that somebody either.

1. Formalize each of these eight statements.
2. Make a preliminary judgment as to whether each of A, B, C, and D provides a satisfactory interpretation of the original statement.
3. Show, both by truth tree and by deduction, that E implies F.
4. Show, both by truth tree and by deduction, that H and E together imply G.
5. Show, as economically as possible, that C implies A.
6. Compare A and B. That is, assuming E, decide by truth tree whether A implies B and whether B implies A. If either of these implications fails, explain why. If either of these implications holds, show that by deduction.
7. Show, both by truth tree and by deduction, that A and E together imply C.
8. Show, both by truth tree and by deduction, that D and H together imply A.
9. Show, both by truth tree and by deduction, that, if A is true, no man I know other than George is such that no man I know is more interesting than he is.
10. Finally, go back to question 2, check your tentative judgments, and make clear how the work you have done answers the questions raised there.

Afterword
On Names and Variables

Variables – symbols used to take the place of noun phrases, verb phrases, sentences, or other locutions, in order to express generality – are, as we have seen, central to the language of logic.[1] Our focus in this book, in introducing the reader to elementary first-order logic, has been on articulating general rules of reasoning. Therefore, from the first we have needed the basic device of the variable for making it visually clear that the rules are intended to apply to all reasoning, regardless of subject matter.

We have selected quite arbitrarily, but according to tradition in the field, letters from various parts of the alphabet, always italicized, to be used in the statement or explication of rules and definitions. We have used 'p', 'q', etc. in place of sentences, 'x', 'y', etc. in place of nouns or noun phrases, 'F', 'G', etc. in place of verbs or verb phrases. We have needed no higher level of generality. In part for that reason, we have been able to avoid the relatively unfamiliar Greek and Cyrillic alphabets.

The letters used as variables do not have referents, but stand as place holders for other symbols that do. They are devices for

---

[1] More than fifty years ago I wrote a Ph.D. thesis on the subject of variables; see Leigh D. Steinhardt, *The Variable in Its Relation to Semantic Problems*, Radcliffe College, 1940. At that time there was an extensive literature dealing with variables. The literature on names and quotation marks is for the most part more recent; see, e.g., Saul Kripke, *Naming and Necessity* (Cambridge, Mass.: Harvard, 1980), first presented as lectures at Princeton in 1970, and Donald Davidson, "Quotation," in *Inquiries into Truth and Interpretation* (Oxford: Clarendon Press, 1985), first published in *Theory and Decision*, 1979. See also W. V. Quine's essay, "The Variable," read at the Boston Logic Colloquium in 1972 and reprinted in *The Ways of the Paradox* (Cambridge, Mass.: Harvard, 1976).

directing attention away from the referents of the symbols that may replace them and drawing attention instead to the logical structure of the sentence forms in which they occur.

It is unfortunate that the use of variables is often explained as involving "systematic ambiguity." (There is a trace of this view in the persistence of the term 'variable', as if something or other "varied" – but what?) 'Systematic ambiguity' is a misnomer: the use of variables need involve no ambiguity at all.

In art and poetry ambiguity is often desirable; not so in logic and science. Within a scientific book or project unambiguous use of basic terms is essential. Accordingly, and especially since the work of Leibniz and Boole, logicians have tried to make their language unambiguous – a major contribution of modern symbolic logic.

Because an ambiguous symbol has two or more different meanings and so can be understood in two or more different ways, it can be confusing to the reader. A variable cannot be confusing in that way, for it has no meaning of its own. We must be very clear, however, about the *scope* of the variables we use – the linguistic space within which they hold their identity – so that there can be no question about the structure of the reasoning. We must be careful also about the *other* words and symbols in our sentence forms, notably '→' and '∧' and '⌐'.

Although I avoided discussion of the use of variables in chapter 1, in fact attempted to avoid discussion of language altogether, I tried to make clear from the beginning that the variables used in explaining the rules of inference – first 'such and such' and 'so and so', then 'p' and 'q' and 'r' – do not have referents. The rules are intended to apply universally. Therefore those variables can be replaced by any sentences that interest us.

I used examples, simple sentences like "Henry is in the kitchen" or "It is raining," both in the text and in the exercises set for the reader. I trusted that the reader would understand that, in doing those exercises, she was not *reasoning*, not in the usual sense of serious inquiry or deliberation before actual decision making. Instead, she is *practicing* in preparation for genuine reasoning, much as a musician practices scales or a tennis player practices strokes before a game. The seeming specificity of the examples is deceptive; it is only a pedagogical tool. It allows the student to use variables without being intimidated by them. The advantage of 'Alice' over 'a' or 'x' is that 'Alice' is more familiar and so less daunting.

All the same, the use of variables is basic to any general discussion. Furthermore, it is not as unfamiliar as many of us tend to think it is. Aristotle introduced the predicate variables 'A' and 'B' ('α' and 'β' in Greek) in the first pages of *Analytica priora*[2] without any apparent need to justify or explain his procedure. In contemporary discussion it often seems artificial not to use variables. When our aim in argument or inquiry is to focus on structure, avoiding the distraction of irrelevant content, it can be artificial (even, occasionally, misleading) to use examples rather than 'p's and 'q's.

Three kinds of variables have been used in this book: *free* variables, *bound* variables, and *pseudonames*. (I am disregarding the use of examples in the text.) It is perhaps more accurate to say that variables have been used in three different ways: with implicit scope, with explicit scope, and with scope limited to a given argument.

Free variables, discussed above, variables not bound by any quantifier or other "binding" mechanism, are used throughout. These variables appear in the explication of rules and in definitions, theorems, and equivalences.

Until this Afterword I have not been at pains to call attention to such usage, primarily because I have found that talking about variables tends to shift the focus of discussion from logic to the philosophy of language, which is beyond the purview of this book. Language is used here as a tool in the service of logic, and logic is designed as a tool to be used in the service of inquiry and rational decision making.

Only sentential free variables ('p's and 'q's) appear in Part I. In Part II, when pronominal bound variables 'x' and 'y' are introduced in explaining quantification theory, free predicate vari-

---
[2] *Analytica priora*, book I, sec. 2, *The Works of Aristotle* translated into English, W. D. Ross, ed. (Oxford: Clarendon Press, 1928), p. 25a.

ables '*F*' and '*G*' are introduced without fanfare. The free sentential variables of Part I continue to be needed, notably in the rules of passage. All these free variables are implicitly bound by universal quantifiers, for the rules and definitions and theorems are applicable universally.

A new dimension is introduced in chapter 4, with the advent of quantification theory. *Bound* variables, explicitly covered by quantifiers, have clearly limited scope, defined by parentheses or brackets. The quantifiers are of two kinds: existential and universal, used to articulate two kinds of general statement, existential and universal. The explicitness of this arrangement makes it possible to deny a universal claim as well as to affirm one, without ambiguity and without overstating one's case.

The only bound variables dealt with in this book are pronominal, '*x*' and '*y*', etc., taking the place of nouns and noun phrases. They are what Whitehead and Russell, following Peano, called "apparent" variables.[3] Because the quantifiers that bind them are explicit and because their scopes – the open sentences in which they occur – are explicit and narrow, it is possible to be quite clear about what it would mean for sentences containing them to be true or false. "Dangling" variables, letters masquerading as bound variables but not properly captured by any quantifier, are stigmatized as unsuccessful, ungrammatical.

Finally, we have used *pseudonames*, single letters chosen from the beginning of the alphabet ('*a*' and '*b*', etc.) and introduced *ad hoc* through the instantiation of quantified statements, within deductions or on truth trees. Pseudonames are not bound by quantifiers, but their scope is nonetheless clearly limited.

Pseudonames are reminiscent of the proper nouns and pronouns of ordinary language. I want to close by commenting on this comparison.

Variables, as we have seen, do not have referents. Real names, the proper nouns of ordinary language, of history and geography, do have referents: things – broadly construed – which can be

---

[3] *PM*, Introduction, 1st ed., p. 17; 2d ed., p. 16. Free variables, which Whitehead and Russell call "real", are discussed on pp. 18 and 19 of the second edition.

found and pointed to or which have been found and pointed to and remembered. Names are assigned to things, including people, as the language grows.

A pseudoname pretends to be a name for a brief space – the duration of an argument – but must be eliminated before a firm conclusion can be reached. Its referent is a fiction, comparable to Othello or Cinderella.

Pseudonames are needed for the two restricted rules of quantification theory, UG (universal generalization) and EI (existential instantiation, or the rule of naming). UG allows us to generalize from a pseudoname, properly derived and so guaranteed not to represent a special case. UG allows no other generalization.

EI allows us to introduce a pseudoname in the course of a deduction, grounded in a statement of existence. But the reasoner is warned immediately that the pseudoname must be new: it cannot be a letter already in use, either in the premises of the argument under discussion, in its proposed conclusion, or elsewhere in the deduction.

Put more generally, this last restriction reminds us that, if language is to function properly, an entity may have many names, but a name may refer to only one entity, at least within a given context of use. Obviously, there are not enough names available to match the vast number of things we want to talk about. Therefore discipline is required, and careful delineation of context.

The warnings about misuse of pseudonames, articulated in chapters 4 and 6, and also related warnings about misuse of descriptive phrases in chapter 9, suggest analogous warnings about misuse of proper nouns. Never generalize thoughtlessly from a single case. Never assign the same name more than once, for fear of confusion. Never assign a name in place of a description, without proof of both existence and uniqueness. Always be as clear as possible about the limits of context of use. If such warnings are heeded they will lead us to be careful in our use of variables in logic and in our use of names in ordinary language; they will help us, accordingly, to think straight.

Index

Symbols

$\forall$	(All): 147, 149, 160
$\wedge$	(And): 30ff. *See* Conjunction
$\rightarrow$	(If): 14ff. *See* Conditional
$\leftrightarrow$	(Iff, if and only if): 45, 52. *See* Biconditional
$=$	(Is, is identical with): 279, 305–6. *See* Identity
$\neq$	(Is not, is other than): 279–80. *See* Otherness
$\sim$	(Not): 35ff. *See* Negation
$\vee$	(Or): 41ff. *See* Disjunction
$\exists$	(Some): 148. *See* Existential
ι	(The, the one and only): 302–6. *See* Description, Iota notation
$\therefore$	(Therefore): 16, 116. *See* Argument

$\forall\exists$ formulas: 236
$\exists\forall$ formulas: 236
$\exists\forall/\forall\exists$ discrimination: 237, 247–51

ADJunction: 32
AND (truth table): 77
AND (truth-tree rule): 86
Affirming the consequent: 18, 96–7
Ambiguity: 31–2, 41, 61, 219, 236, 248–9, 303, 319–22, 331–2
Anderson, John M.: 21 fn 3
Antecedent: 14–5
Antilogism: 124
Argument: 3, 13, 73

Aristotelian tradition: 147, 155–6, 159 fn 7, 161, 170, 218, 229, 285
Aristotle: 38, 113, 218, 333
Assessment of validity: 120–7, 206
Auxiliary definitions: 231, 234
Auxiliary premises: 6, 22, 26, 235
Auxiliary rules of inference: 24, 55–6, 68–9, 173

Axiom of self-identity (BL, Butler's law): 281, 318-9

BL (Butler's law, or axiom of self-identity): 281
BT (Butler's law for truth trees): 290
Bergmann, Merrie: 21 fn 3
Biconditional: 45, 52, 81
Bound variables: 149, 333-4
Bizet, Georges: 115
Blumberg, Albert E.: 21 fn 3
Bohnert, Herbert: 78 fn 2
Boole, George: 113, 332
Butler, Joseph: 278

CHAIN rule: 22-25, 55
CP (Conditional Proof): 5, 19
CV (Change of Variable): 151, 167
Carroll, Lewis: 220
Citation: 54, 62-4, 128
Completeness: 13-4
Conclusion: 3, 11-12. *See also* Summary conclusion
Conditional: 14 ff. *See also* Corresponding conditional
Conjunction: 30-5
Connectives: 45, 52, 73, 111-5. *See also* Main connectives, Non-truth-functional connectives
Consequent: 14-5
Consistency: 86-7, 92-3, 98, 123-5
Contradiction: 12, 36, 290, 292. *See also* (law of) Non-contradiction

Contraposition: 67, 82
Corresponding conditional: 5, 19-20, 99
Corresponding fallacy: 18, 84, 127
Counterclaim: 92-3, 98, 124
Counterexample: 73, 84, 98, 123
Counterfactual: 114-5
Cross flagging: 237-40, 269
Cross reference: 149, 151, 175, 234, 243-46, 295
Cyclical flagging: 240

DILemma: 41-2, 51
DN (Double Negation): 37-9. *See also* truth table for NOT: 77-8
Dangling variables: 149, 174, 244
Davidson, Donald: 331 fn 1
Deduction: 4-5. *See also* Presenting deductions
De Morgan, Augustus: 81, 113
De Morgan's laws: 54, 81
Denying the antecedent: 84, 127
Description: 302-6
Dilemma: 5-6, 41-3, 51
Disambiguation: 116-7, 249-51
Discharging premises: 5-6, 21, 29-31, 36, 42, 64
Disjunction: 41-3
Double flagging: 163, 166, 180, 210, 239-40

EG (Existential Generalization): 154
EI (Existential Instantiation, or Rule of Naming): 157, 187
ELIMination: 56
ET (truth-tree rule for existentials): 189
Eliot, George (Mary Ann Evans): 278, 293 ff.
Emerson, Ralph Waldo: 240 fn 3
Enthymeme: 126
Equivalence: 39 fn 6, 52-4, 245. *See also* Proofs of equivalence
Equivalences: 54. *See also* Lists of equivalences
Equivalences for Negation: 93, 95, 168-70
Excluded Middle: 26, 49-51, 100-1
Exclusive "or": 41, 44, 89-90, 116
Existential import: 218-9, 315-6, 320
Existentials: 153-55, 187-9

Fallacy: 18, 96-7. *See also* Affirming the consequent, Cross flagging, Cyclical flagging, Denying the antecedent, Double flagging, and Illicit repetition. *See also:* Identifying fallacies
Flag: 157, 159
Finished deduction: 157, 163-4

Formalization: 1-2, 15, 17, 111-4, 117-8, 146, 154. In predicate logic: 173-6, 203-5, 208. In relational logic: 230-4, 241-4, 248-51, 257. With identity: 284, 288-91, 302-6, 319-21. *See also:* Iota notation
Free variables: 149, 231, 279, 333

Gaps in argument: 121-2
Generality: 145-7
Generalizing: 160, 238
Gentzen, Gerhard: 42 fn 7
Goodman, Nelson: 115 fn 2
Grouping: 61, 88, 113

Hypothetical reasoning: 11-2
Hypothetical syllogism: 22. *See also* Chain Rule

IF (equivalence): 48, 75
IF (truth table): 76
IF (truth-tree rule): 90
IFF (equivalences): 82-3
IFF (truth table): 83
IFF (truth-tree rule): 91-2
INTerchange of equivalents: 52
Identifying fallacies: 268-9
Identity: 26, 277 ff., 305-6. *See also* Self-identity
Illicit generalization: 160
Illicit repetition: 29-30, 33, 120, 269
Imagination: 2, 3
Implicit negation: 94-5, 190

Incomplete trees: 196–7
Inconsistency: 12, 16, 86, 125, 277, 290
Inconsistent premises: 216–7
Indirect proof: 4, 6, 12, 36, 92, 98, 101
Induction: 12
Infinite trees: 263–6
Instances: 46–7, 146, 188–9
Interchange of equivalents: 39 fn 6, 52 fn 8, 52–4, 151, 167
Internalizing negation: 93–4, 170, 190
Invalidity: 4, 6–7, 124
Iota notation: 302–6
Ishiguru, Hide: 279 fn 2

Jeffrey, Richard C.: 85, 94, 115 fn 2, 240 fn 3
Johnson, Charles: 176 fn 13
Johnson, Lyndon: 321
Johnson, W.E.: 161 fn 9
Johnstone, Henry W.: 21 fn 3
Justification: 23, 30, 63–4

Kennedy, John: 321
Kripke, Saul: 331 fn 1

LL (Leibniz's law): 279
LL' (Leibniz's law in the negative): 280
LT (Leibniz's law for truth trees): 292
Ladd-Franklin, Christine: 124–5
Lambert, Karel: 153 fn 4
Leibniz, Gottfried Wilhelm von: 113, 279 fn 2, 332
Leutz, Emanuel Gottlieb: 92
Lincoln, Abraham: 248
Lists of equivalences: 67, 93, 103–4, 133–4, 175–6, 182, 243, 253
Lists of theorems: 65, 134, 324–5

MP (Modus Ponens): 16
MT (Modus Tollens): 55, 146
Main connectives: 34, 54, 117 ff.
Metalogic: 1, 39 fn 6, 52 fn 8, 101–2, 110, 151 fn 3, 266
Metaphysical assumptions: 35, 153, 199–200, 218–9
Middleton, Nancy: 20 fn 2
Moor, James: 21 fn 3

NAND (equivalence): 81
NAND (truth table): 78
NAND (truth-tree rule): 94
NCONJ (NonCONJunction): 56
NIF (equivalence): 79
NIF (truth table): 79
NIF (truth-tree rule): 94
NIFF (equivalences): 83
NIFF (truth table): 83
NIFF (truth-tree rule): 94
NOR (equivalence): 80
NOR (truth table): 80
NOR (truth-tree rule): 94
NOT (truth table): 77–8
Names: 149, 157–8, 187–8, 195, 334–5
Naming, rule of: 157. *See also* EI

Natural deduction: 4
Negation: 6, 35–41, 77–83, 167–70, 190–1, 303, 319–22. *See also* Equivalences for negation and Internalizing Negation
Nelson, Jack: 21 fn 3
Nested quantifiers: 148 fn 2, 208 ff., 229, 231–2, 250, 257. *See also:* Order of nested quantifiers
Noncontradiction: 35–6, 153
Nonempty universe: 153, 153 fn 4, 165, 199–200
Non-truth-functional connectives: 114–6
Number One: 282, 284–90

OR (truth table): 80
OR (truth-tree rule): 86
Only: 16, 288, 310
Only if: 16, 114, 310
Open sentence: 148–9, 148 fn 1, fn 2, 231
Order of nested quantifiers: 236–7, 247, 250
Otherness: 236, 255, 277–9, 282–3

PREM ("rule" of PREMises): 20, 22
Parentheses: 34, 61, 116, 120, 135
Particular syllogism: 159
Peano, Giuseppe: 334
Peirce, Charles Sanders: 113, 124 fn 7

Premises: 2–3, 11, 13, 20
Premises in play: 20, 32
Prenex form: 172, 213, 247, 283
Presenting deductions: 62–5
Proofs of equivalence: 53–4, 75–6, 79
Pseudonames: 149, 157–8, 166, 188–9, 333–5
Pure form: 172, 212, 247, 283

QED (*quod erat demonstrandum*): 23, 60, 163 fn 11
Q-INT (Quantificational INTerchange): 151, 167
QN (Quantificational Negation): 168
QN' (Aristotelian QN): 170
Quantification: 146
Quantificational Negation: 167–70
Quantifiers: 147–9. *See also* Scope of quantifiers
Quine, Willard Van Orman: 21 fn 3, 76 fn 1, 85, 115–6, 125, 148 fn 2, 157, 159 fn 8, 173, 176 fn 14, 304, 324, 331 fn 1

REDuctio ad absurdum: 36. *See also* Indirect Proof
REPEAT: 26
Reasoning: 1–2, 5–6
Repetition, fallacy of illicit: 29–30, 33, 120, 269
Restriction on REPEAT: 28–30, 32–3

Restrictions on EI and UG: 157, 160, 163, 229, 239–40, 251–2
Roosevelt, Eleanor: 115
Rules of Inference: 11, 13–4, 20, 45, 54, 150, 154, 279–81. Lists of: 56–8, 130–1, 180–2, 252–3, 280–1, 298–9
Rules of Passage: 172–9, 247. List of: 175
Russell, Bertrand: 148 fn 1, 302, 304, 307, 319, 324, 344

Sample arguments: 17, 24–5, 55
SIMPlification: 30
Sacco, Niccola: 315–6
Scope of quantifiers: 149, 172–6, 192, 203–4, 243 ff.
Self-identity: 281, 319. *See also* BL, Butler's law
Singular statements: 154–5, 156 fn 5, 285–6, 285 fn 3
Sommers, Fred: 156 fn 5, 285 fn 3
Star system: 19, 21–4, 30, 59–60, 128
Statement: 3, 46, 73–4, 87, 99–102, 115, 117–8
Steinhardt, Leigh D.: 331 fn 1
Stoics: 147, 229
Strategy in assessment: 120–1, 125, 312
Strategy in deduction: 5–6, 25–8, 37, 43, 50, 127–8, 179–80, 235, 249, 312
Strategy in formalizing: 175, 203–5, 231, 234, 244–6, 251, 258
Structure of argument: 2, 13, 17, 111–4, 117, 145, 174
Structure of statements: 32, 34, 117
Summary conclusion: 5, 19–21
Suppositions: 2–3
Syllogisms: 124, 147, 155, 193
Symbols: 2, 15, 17, 73, 111, 332–3

THEOREM: 45–8, 68
THINning: 41, 43–4
Tautology: 46, 96, 99–100
Telescoping: 116, 119
Testing for validity: 4, 73–4, 84–5, 92–3, 98, 100–1
Theorems: 3–4, 26, 37, 39–41, 46–7, 99–100, 128, 315, 318–9. Lists of: 65, 134–5, 324–5
Truth: 1, 102
Truth functions: 74, 145
Truth in packaging: 23, 59
Truth preservation: 1, 11–3, 44
Truth tables: 74–85, 103
Truth-tree rules: 86, 90–1, 93–4, 189, 290–3. Lists of: 103, 189, 219–20
Truth trees: 4, 85 ff., 187, 257, 290 ff.
Truth value: 73

UG (Universal Generalization): 160
UI (Universal Instantiation): 150
UT (truth-tree rule for universals): 189
Unfinished deduction: 163–4
Unfinished trees: 196, 200, 259
Unity: 282, 284, 289, 302, 318–9
Universals: 147, 154

Unless: 114

Validity: 1, 3–4, 7, 13, 46–7, 73, 99
Vanzetti, Bartolomeo: 315–6
Variables: 17, 148, 331–3
Verdi, Giuseppe: 115
von Fraassen, Bas C.: 153 fn 4

Whistler, James McNeill: 115 fn 2
Whitehead, Alfred North: 148 fn 1, 302, 307, 319, 344

Preferences

Edited by Christoph Fehige and Ulla Wessels

1998. 23 × 15,5 cm. LXIX, 568 pages. With numerous figures.
Cloth. ISBN 3-11-015910-4
Paperback. ISBN 3-11-015007-7
(Perspektiven der Analytischen Philosophie/
Perspectives in Analytical Philosophy, Vol. 19)

Preferences explores the concept and the role of preferences in practical reasoning. What, the contributors ask, is the role of preferences, desires, and wishes in rational decisions? What is their role in ethics, and in between: can preferences *ground* ethics in rationality?

"Outstanding intellects grapple with a central, difficult topic. Editors and authors together sustain a marvelous tone of careful, clear-headed, and understandable argument ... Delivers great pleasure and great benefit, even to non-experts."
Nuel Belnap, Department of Philosophy, University of Pittsburgh

"Illuminating ... Conveys a vivid impression of current analytic value-theory and its controversies."
Georg Henrik von Wright, Department of Philosophy, University of Helsinki

Actions, Norms, Values
Discussions with Georg Henrik von Wright

Edited by Georg Meggle
Assisted by Andreas Wojcik

1998. 23 × 15,5 cm. XIX, 370 pages. Cloth. ISBN 3-11-015484-6
(Perspektiven der Analytischen Philosophie/
Perspectives in Analytical Philosophy, Vol. 21)

"Actions, Norms, Values" are the main topics in the philosophical work of Georg Henrik von Wright (*1916). This volume presents the proceedings of the international von Wright-conference at the Center for Interdisciplinary Studies in Bielefeld (Germany), 1996. Contributions are by philosophers, law theoreticians and political scientists.

Walter de Gruyter  **Berlin · New York**